ANIMAL BIOCHEMISTRY

ANIMAL BIOCHEMISTRY

Edited by

Prof. Dr. Nandkumar Gaikwad

Delve
PUBLISHING

www.delvepublishing.com

Animal Biochemistry

Prof. Dr. Nandkumar Gaikwad

Delve Publishing
4164 Lakeshore Road
Burlington, ON L7L 1A4
Canada
www.delvepublishing.com
Email: orders@arclereducation.com

ISBN: 978-1-77956-360-6 (Hardcover)

ABOUT THE EDITOR

Dr. Nandkumar Gaikwad (b. 1965) is presently serving as Associate Dean and Professor of Veterinary Biochemistry, College of Veterinary and Animal Sciences, Udgir under Maharashtra Animal and Fishery Science University Nagpur (M.S.). He did his B. V.Sc & A. H. in 1989 and M. V. Sc. (Vet. Physiology) in1991 from College of Veterinary and Animal Scieces, Parbhani (MAU, Parbhani), Ph. D. (Physiology and Biochemistry) in 2005 from Maharashtra Animal and Fishery Science University Nagpur. He started his career as Assistant Professor (Vet. Physiology and Biochemistry) in 1993 and as a Associate Professor in 2009. His field of specialization is Veterinary Biochemistry and his area of Masters research on "Hematological and some blood biochemical constituents in Jersey X Red Kandhari cattle with reference to age and sex". His area of Doctoral research on "Effect of water deprivation in Osmanabadi goats of Marathwada region". During his total 32 year's tenure of services, under his guidance 02 students completed M.V.Sc., 02 students of Ph. D. as Co-guide and he acted as co-guide for 67 students in various disciplines of Veterinary education .

He has published 57 research papers in international and national reputed journals; 05 practical manuals for under graduation/post graduation/Ph.D. courses;. He had participated and presented research papers more than 10 International, National, State Seminar, workshops and symposiums. Besides these activities Radio talks (11), guest Lectures in scientist training & farmer training programmes (27). In addition to above he also written more than 25 popular articles in reputed regional language periodicals and daily news papers.

Additionally during 32 years of service period he has handled the responsibilities as Officer Examination Cell, Chairman Gymkhana, NSS Program Officer, Hostel Warden, MAFSU Sub-Center In-charge, Head of Department of Animal Genetics and Breeding, Veterinary Physiology, Anatomy and Histology, etc .

TABLE OF CONTENTS

LIST OF FIGURES

ABSTRACT

"Animal Biochemistry" offers a comprehensive overview of the biochemical processes that are the foundation of animal physiology. The book provides insights into how animals utilize various molecules and pathways to maintain their physiological functions by delving into the metabolism of carbohydrates, lipids, proteins, and nucleic acids. This is accomplished by examining the metabolism of these substances. As an additional point of interest, the book sheds light on how animals coordinate their metabolic responses in order to adapt to changing environmental conditions by examining the regulation of metabolic pathways. In addition this book also helpful for clinical diagnosis of different diseases on the basis of biochemical parameters. Students and researchers who are interested in understanding the biochemical basis of animal physiology will find "Animal Biochemistry" to be an invaluable resource because of the book's in-depth examination of the aforementioned topics.

PREFACE

Animal Biochemistry is an intriguing discipline that explores the biochemical mechanisms taking place in the bodies of animals. This book provides a thorough and detailed explanation of the complex mechanisms of biochemistry within the framework of animal physiology. Through the examination of chemical reactions and pathways occurring within living organisms, we can develop a more profound understanding and admiration for the intricate and esthetically pleasing aspects of the natural world.

This book has a wide-ranging focus, encompassing subjects such as metabolism, hormone regulation, enzyme function, and cellular signaling pathways in animals. The book seeks to enhance our comprehension of the molecular-level functioning of animals by investigating these fundamental biochemical processes.

This book is significant because it has the capacity to connect the fields of biochemistry and animal physiology. The book provides valuable insights into the molecular mechanisms that govern physiological processes in animals, which can be applied to research in fields such as veterinary medicine, animal nutrition, and evolutionary biology.

The information contained in this book is urgently required due to the escalating intricacy of animal biochemistry research. With the advancement of our knowledge about biochemical processes in animals, there is an increasing need for a comprehensive resource that consolidates the most recent discoveries in the field. This book aims to meet this requirement by offering a state-of-the-art summary of the molecular foundation of animal physiology.

The book's uniqueness and value lie in its interdisciplinary approach to the examination of animal biochemistry. The book provides a comprehensive view of the biochemical processes that regulate animal life by incorporating ideas from biochemistry, physiology, and molecular biology. This distinctive methodology distinguishes the book from other publications in the same field, rendering it a valuable asset for both researchers and students.

This book will appeal to a diverse audience, encompassing biochemists, physiologists, veterinarians, as well as biology and related field students. This book will provide valuable insights for researchers studying animal nutrition, metabolism, and hormonal regulation.

Through the act of reading this book, the reader will acquire a more profound comprehension of the molecular mechanisms that propel physiological processes in animals. The book offers an in-depth examination of crucial biochemical pathways and signaling cascades, elucidating the mechanisms by which these processes are controlled in various animal species. The book provides the reader with the necessary knowledge and tools to conduct further research in the field of animal biochemistry by incorporating advanced research and thorough analysis.

Ultimately, "Animal Biochemistry" is an invaluable resource for individuals seeking to acquire a more profound comprehension of the biochemical mechanisms that regulate animal physiology. This book presents an extensive analysis of recent research in the field, offering a distinct viewpoint on the molecular mechanisms that drive physiological processes in animals. If you are involved in animal biochemistry as a researcher, student, or practitioner, this book will undoubtedly offer valuable insights and stimulate further investigation into this captivating field of study.

Chapter **1**

Introduction to Animal Biochemistry

Contents

Biochemistry refers to the scientific investigation of chemicals and chemical reactions occurring in plants, animals, and microorganisms (Zaitsev et al., 2020). It utilizes techniques from analytical, organic, and physical chemistry to investigate the chemical processes of living organisms. Metabolism encompasses all the chemical reactions that take place within an organism, including the conversion of substances into energy and the synthesis of complex molecules necessary for survival. Enzymes are organic catalysts that rely on the cellular genetic machinery for their existence, making these modifications essential. Biochemistry is indispensable for comprehending chemical alterations related to disease, medication efficacy, nutrition, genetics, and agriculture (Zaitsev et al., 2020).

Physiological chemistry and biological chemistry are interchangeable terms for the field of biochemistry, which encompasses the study of large molecules like proteins and nucleic acids, as well as molecular biology. This chapter will explore key facets of biochemistry, encompassing: an overview of animal biochemistry (Figure 1.1), the significance of studying animal biochemistry, the historical advancements in animal biochemistry, the fundamental principles of biochemistry, the methodologies employed in animal biochemistry, the practical applications of animal biochemistry, and the prospective directions in animal biochemistry.

Figure 1.1. Animal biochemistry.

Source: Usman (2024).

1.1. OVERVIEW OF ANIMAL BIOCHEMISTRY

In the field of veterinary science, biochemistry plays a vital role. To diagnose uncontrolled diseases in the health of both wild and captive animals, it is necessary to analyze blood hematological and biochemical indicators. Serum biochemistry is a valuable method for assessing the health of both domestic and wild animals, and hematological traits play a vital role in this evaluation. The evaluation of hematological and biochemical parameters is dependent on the species being studied, as it influences the nature of their changes. The utilization of animal blood as a diagnostic tool for disease pathology is crucial in the field of wildlife management. This is because herbivores are believed to have economic benefits for humans and play a crucial role in ecology (Daly et al., 2020).

1.1.1. Application of Clinical Biochemistry in Wildlife Treatment

Clinical biochemistry, as defined by Kaneko (2000), refers to the examination of blood plasma, or serum, for various substances such as hormones, enzymes, and substrates. This analysis is not only used for diagnosing and monitoring wildlife diseases, but also for studying other body fluids. Tests also include the analysis of ascitic, cerebrospinal, and urine fluids obtained from wildlife. Various tests are employed to identify a specific clinical condition due to the presence of multiple factors that affect the commonly requested analytes (Kaneko, 2000). According to this theory, a meticulously chosen battery of tests can be conducted to provide specific information about a variety of conditions that are pertinent to the management of wildlife. In addition, biochemistry tests are used in combination with complete hematology tests to identify specific disease patterns, as the evaluation of both sets of results is necessary (Kaneko, 2000).

1.1.2. The Fundamental Test Panel in the Management of Wildlife

Performing a basic panel of tests is feasible in multiple veterinary laboratories. Each panel generally represents the minimum amount of research required. Develop a standardized panel for the analysis of small animals that encompasses the measurement of albumin, creatinine, urea, total protein, globulin, and alkaline phosphatase, as outlined in Kaneko's (2000) study.

Additionally, it is possible to conduct tests to measure bilirubin levels. This panel could be altered for the examination of different animal species. According to Kaneko (2000), it is believed that gamma-glutamyl transferase (GGT) and/or glutamate dehydrogenase (GDH) are superior liver enzyme tests for agricultural animals in terms of their diagnostic value. However, when studying athletic animals, it is believed that muscle enzymes (CK and AST) are more suitable for examination.

1.1.3. Animal Clinical Biochemistry

*** As clinical biochemistry is helpful in diagnosis of various diseases so it's the importance of each parameter for diagnosis is tobe mentioned along with normal range values of each biochemical parameter in different species of animals in tabular form for ready reference and understanding.*

Animal biochemistry is the scientific investigation of various chemical processes that take place within the body of an animal. This entails a more profound understanding of metabolism and its functioning in both the well-being and illness of animals. The comprehension of how the chemical composition of biological molecules is connected to their purpose is founded on the field of biochemistry, which plays a vital role in metabolism and function (Washington & Van Hoosier, 2012). Understanding and analyzing chemical factors is essential in the field of animal husbandry. The application of biochemistry techniques facilitates the growth and cultivation of domestic animals. Veterinarians employ a diverse range of biochemical instruments to diagnose diseases in animals. Modern technologies such as dynamic tensiometry (DT) are used in the field of cow husbandry (Washington & Van Hoosier, 2012). Figure 1.2 depicts the clinical biochemistry.

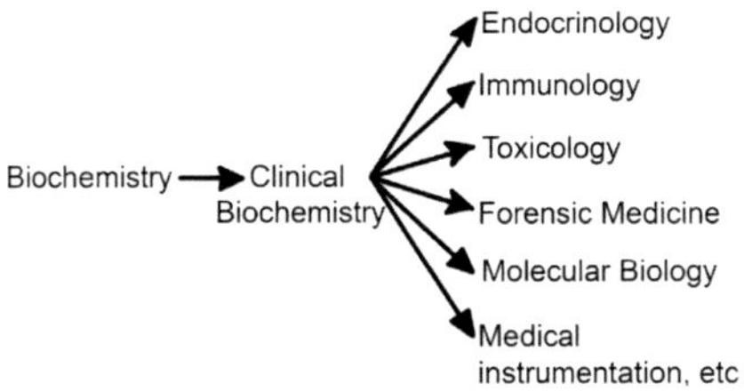

Figure 1.2.*Clinical biochemistry is a clinical-specific laboratory science.*

Source: Jaypee Digital | Clinical Biochemistry (2015).

In order to conduct a thorough assessment of animal health in cattle farming, blood serum DT and PCR analysis of specific markers are employed. Hematological parameters such as total protein and its fractions, enzymes, low molecular metabolites like glucose, lipids, and bilirubin, as well as cations and anions, are also employed for analysis. Research on cell-free nucleic acids present in the bloodstream is specifically employed to examine and assess conditions such as cancer, viral and chronic diseases, and pregnancy (Washington & Van Hoosier, 2012). Conventional techniques in molecular biology, such as next-generation sequencing and DNA amplification, can be employed to detect and characterize this cell-free DNA. Furthermore, advanced techniques such as digital PCR offer enhanced precision in quantifying copy number variations, which play a vital role in detecting any chromosomal abnormalities in the fetus (Washington & Van Hoosier, 2012).

1.1.4. Utilizing Biochemistry in Animal Reproduction

Cattle farmers consider fertility to be a desirable characteristic in animals raised for cattle. Fertility can be influenced by various factors, such as genetics, health, and environmental factors like management techniques. By specifically targeting key stages in the reproductive process, biochemistry has been employed to manipulate fertility in different animal species, such as cattle. The processes involved in reproduction encompass ovulation, fertilization, implantation, gestation, and gamete production (Daly et al., 2020). The food and agriculture sector can now effectively meet the needs of a growing human population by using biochemical techniques to manipulate fertility, resulting in the production of milk and meat.

1.1.5. Application of Biochemistry to Improve Livestock

The utilization of genome or gene-editing (Figure 1.3) technologies is increasingly being investigated.

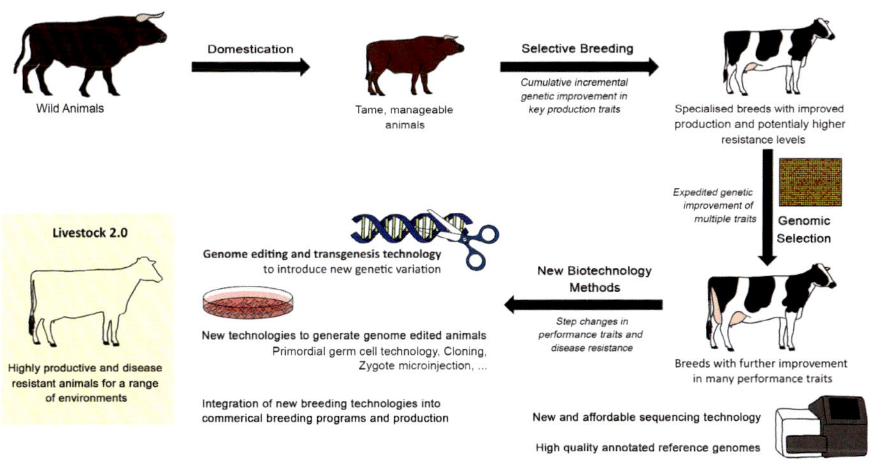

Figure 1.3. *Pathways to genome editing for fitter, healthier, and more productive farmed animals.*

Source: Tait-Burkard et al. (2018).

Homologous recombination is a process that allows for precise modification of a specific bovine genomic locus in somatic cells, resulting in the elimination or alteration of certain genes. The application of this gene-targeting technique has primarily been demonstrated in mouse embryonic stem cells, serving as a proof of concept.

The advancement of designer nucleases, such as zinc-finger nucleases and transcription activator-like effector nucleases (TALENs), has facilitated more efficient and uncomplicated genome engineering in animals. These designer nucleases are known as CRISPR associated protein 9 (CRISPR/Cas9), which are clustered regularly interspaced short palindromic repeats (CRISPR) (Crowe et al., 2018). As a consequence, transgenic animals are created, possessing the capacity to generate meat with an enhanced nutritional composition, milk containing beneficial medicinal elements, or animals that exhibit heightened resistance to diseases.

1.2. IMPORTANCE OF STUDYING ANIMAL BIOCHEMISTRY

Biochemistry is the scientific discipline that investigates the chemical substances and processes occurring in living organisms. The primary focus of animal biochemistry is the examination of chemicals, their interactions with enzymes, and their effects on the entire body of an animal. The scope of this study is continuously growing, and it contributes to the comprehension of the structure and impacts of various substances within an animal's body. The primary subjects of biochemistry include molecules, lipids, metabolisms, and various other elements (Soetan et al., 2010). To gain a deeper understanding of the importance of biochemistry in animal life, it is helpful to examine the following five distinct branches:

1.2.1. BIOMOLECULES IN ANIMAL BIOCHEMISTRY

The field of biomolecules is of utmost importance in the study of animal biochemistry. The molecules are divided into four distinct classes, which are described as follows:

1. **Carbohydrate:** A monomer is a minuscule molecule that chemically bonds with other monomers of similar size to create a polymer. The process of dehydration synthesis is involved in the formation of carbohydrates, which are macromolecules composed of monosaccharides.

2. **Lipid:** These consist of glycerol molecules that are chemically bonded together. A triglyceride is composed of one molecule of glycerol and three molecules of fatty acid. These lipids are also used as an intravenous infusion in various medications (Soetan et al., 2010). Some medications contain liposomes, which act as drug carriers.
3. **Protein:**This consists mainly of amino acids and various other molecular groups. Dehydration arises during the process of synthesizing amino acids into polypeptides, which are alternatively referred to as proteins.

The side-chain R group and carboxyl group are the additional components used in constructing the molecule. Proteins consist of the carboxyl group, side-chain R, and amino acids (Soetan et al., 2010).

4. **Nucleic Acid:** Deoxyribonucleic acid is a derivative of nucleic acid. Genetic information is stored by the entire organism using DNA. The field of animal biochemistry allows researchers to analyze genetic information for the purpose of developing potent medications and therapies for chronic diseases.

A carbohydrate is the most abundant biomolecule in the universe. Glucose, fructose, and deoxyribose are the sources from which they are derived, with deoxyribose being the most abundant (Soetan et al., 2010).

1.2.2. Animal Biosynthesis

This refers to the process by which animals produce complex organic compounds, such as proteins and carbohydrates, through biochemical reactions within their cells.

This branch of animal biochemistry focuses on the following:

1. **Biosynthesis of Glutamate and Aspartate:**The α-keto acid is the main molecule that carries out a fundamental transamination process and generates glutamate and aspartate. Aspartate plays a crucial role in the urea cycle and serves as the precursor for ornithine. According to researchers, aspartate is a compound that is derived from asparagine (Soetan et al., 2010).
2. **Biosynthesis of Alanine:** Alanine is primarily derived from animal tissues present in muscles. The process of alanine transmission operates in both directions, with alanine being transported from pyruvic acid to the liver, and glucose being transported from the liver to cells. When the liver receives an adequate amount of alanine and eliminates nitrogen, the carbon structures undergo inversion.

3. **Biosynthesis of Tyrosine:**Phenylalanine hydroxylation is utilized by animal cells to produce tyrosine. The synthesis of neurotransmitters and other vital brain chemicals is crucial for the survival of the organism. Around 50% of this reaction involves the utilization of oxygen and phenylalanine. If the organism consumes a diet high in tyrosine, its phenylalanine requirement will decrease to 40–60% (Soetan et al., 2010).
4. **Biosynthesis of Ornithine and Proline:** Living organisms necessitate arginine for their survival. When you feed your carnivorous pet white meat and treats containing nuts, its urea cycle replaces the need for this carbamoyl phosphate acceptor. Herbivores and omnivores have distinct nutritional requirements. Glutamate transamination is necessary for the synthesis of ornithine (Soetan et al., 2010).

1.2.3. Animal Biochemistry Synthesis

The subdivisions within this area of animal biochemistry are as follows:

1. **Lipid Synthesis:**The Krebs cycle consists of three main constituents: lipids, proteins, and carbohydrates. The Krebs cycle activity at the main gate necessitates the presence of the co-enzyme A Acetyl. The three components mentioned above fulfill each other's requirements and facilitate the synthesis necessary to achieve the desired effects. Carbohydrates can facilitate lipid production in the absence of protein. Pyruvic acid is the byproduct obtained from this process that produces glycerol.
2. **Protein Synthesis:**Amino acids combine to form polypeptide molecules. The cell utilizes these amino acids in situations pertaining to DNA and RNA. Each individual cell possesses the ability to synthesize its own proteins and acquire the necessary energy. DNA and RNA, commonly referred to as genetic agents, play a crucial role in regulating the synthesis of proteins, making them indispensable to this biological process (Soetan et al., 2010).

1.2.4. Amino Acid Metabolism

Amino acids are essential for amino acid metabolism, which accounts for approximately 75% of the total utilization of amino acids. Protein consumption is essential for the metabolism of amino acids. It is highly recommended to provide lean protein for your cat. Cats' digestive tracts are unable to process saturated protein and fat from large animals over a long period of time. Dogs typically consume and ingest saturated protein. Amino acids serve as a source of nitrogen for various cellular components and other biochemicals. Amino acid synthesis is crucial for the formation of hormones, the alkaline structure of DNA and RNA, and the production and transmission of neurotransmitters (Soetan et al., 2010).

- **Metabolic Pathways of Amino Acids:** This pathway comprises four sequential steps:
  - The liver synthesizes amino acids to facilitate the production of amino acids during protein digestion.
  - Extracting necessary nitrogen from amino acids.
  - Amino acid catabolism occurs through the acid and urea cycles.
  - The final stage in the process of producing amino acids through protein synthesis.

1.2.5. Carbohydrate Metabolism

Glucose serves as the primary energy source for fetuses, both in humans and animals. The primary mode of sustenance for organisms and cells is the utilization of this agent as a source of energy. Glucose, a ubiquitous energy source, is synthesized within living organisms from a diverse range of food sources. Glucose is transported into the body through blood vessels and subsequently undergoes conversion in the liver.

- **Metabolism Classifications:** The anabolic pathway, also referred to as protein synthesis, is responsible for shaping the form of organs and bone structure.

The respiration chain and oxidative phosphorylation processes in the catabolic route produce high-energy phosphate. The amphibolic pathway acts as a connection and enables various actions in both anabolic and catabolic pathways (Soetan et al., 2010).

1.3. EVOLUTION OF ANIMAL BIOCHEMISTRY THROUGHOUT HISTORY

Biochemistry has been commonly known by that name since approximately 1900. However, its origins extend much further; its early origins are closely connected to the origins of physiology and chemistry.

Significant advancements in the field of biochemistry have primarily focused on establishing a strong chemical foundation for understanding biological processes.

Chemistry needed to dissociate itself from immediate practical requirements in order to establish itself as a fundamental science, prior to making a significant impact on agriculture and medicine (Michal & Schomburg, 2012). During the period spanning approximately 1650 to 1780, significant scientific advancements were made, starting with the contributions of Robert Boyle and culminating with the groundbreaking work of Antoine-Laurent Lavoisier, who is considered the pioneer of modern chemistry. Boyle questioned the fundamental principles of the prevailing chemical theory of his time and argued that the primary objective of chemistry should be to determine the composition of a substance. John Mayow, a contemporary of his, observed the fundamental resemblance between an animal's respiration and the process of oxidation, which involves the combustion of organic substances in the atmosphere. Lavoisier conducted fundamental research on the subject and discovered the true nature of the process by demonstrating the quantitative similarities between chemical oxidation and the respiratory process (Michal & Schomburg, 2012).

Photosynthesis, a biological phenomenon, piqued the curiosity of chemists in the late 18th century. The discovery that photosynthesis is essentially the reverse of respiration, thanks to the collaborative work of Joseph Priestley, Jan Ingenhousz, and Jean Senebier, marked a significant milestone in the development of biochemical theory.

Prior to the rapid advancement of biochemistry, the development of structured organic chemistry, which was a significant achievement in the 19th century, had to occur. A living organism is composed of a vast number of unique chemical constituents (Michal and Schomburg 201). An essential concern in the field of biochemistry is elucidating the molecular transformations that occur in these compounds within a living cell. Prior to studying the molecular structures of organic materials in live cells, it is essential to have a clear understanding of the cellular processes involved in their synthesis and degradation (Michal & Schomburg, 2012).

Science exhibits a lack of distinct boundaries, and the divisions between physical and organic chemistry and biochemistry have consistently demonstrated substantial overlap (Michal & Schomburg, 2012). Biochemistry has adopted the principles and methods of physical and organic chemistry to study physiological problems. The misconception that the changes occurring in living organisms were not subject to the same chemical and physical laws as inanimate objects hindered progress in understanding these "vital" phenomena in terms of common chemistry or physics initially. The vitalists, who believed that conventional chemical processes were incapable of producing natural products created by living organisms, embraced this perspective (Michal & Schomburg, 2012).

In 1828, Friedrich Wöhler conducted a laboratory synthesis of urea, which was the first organic substance. This achievement weakened the beliefs of vitalists, although it did not completely disprove their theories. They adopted defensive strategies and argued that urea was a byproduct of breakdown rather than synthesis. Due to the successful synthesis of various natural compounds by organic chemists, the vitalists were forced to make additional retreats. In modern biochemistry, it is widely accepted that the same chemical principles that govern non-living objects also apply to living cells.

The progress of the new science was influenced by the practical needs of people, while also being hindered by an inappropriate level of respect for living things.

In the 19th century, organic and physical chemistry developed a substantial theoretical foundation. The practical needs of physicians, pharmacists, and agriculturalists prompted the application of these chemical discoveries to various urgent issues. Unlike animals, green plants only need carbon dioxide, water, mineral salts, and sunlight for their growth. The acquisition of minerals is contingent upon the extraction from the soil, while the productivity of the soil is reliant on its capacity to provide the plants with these indispensable nutrients. However, the soil becomes exhausted of these substances due to the continuous removal of crops, thus necessitating the use of fertilizers. Liebig emphasized that the chemical analysis of plants could be used as a reliable indicator of the substances that ought to be included in fertilizers (Michal & Schomburg, 2012).

1.4. FUNDAMENTAL PRINCIPLES OF BIOCHEMISTRY

Biochemistry has a historical background spanning 400 years. The term "biochemistry" was first used in 1882, but it is generally attributed to German scientist Carl Neuberg who coined the phrase in 1903. Biochemistry is the scientific discipline that investigates the chemical reactions occurring within living organisms. Biochemistry governs all living organisms and their activities. The intricate intricacy of life arises from biochemical processes that govern the transmission of information through biochemical signaling and the transfer of chemical energy through metabolism (Voet et al., 2008).

Biochemistry primarily focuses on the structures and functions of various biological components such as proteins, lipids, carbohydrates, nucleic acids, and other biomolecules. However, there is a growing emphasis on studying processes rather than individual molecules.

The main focus of pure biochemistry today is to comprehend how biological molecules contribute to the processes that take place within living cells. This has significant implications for our understanding and capacity to study complete organisms. Polymers are large and intricate compounds composed of repeating subunits called monomers. They are just one type of the many diverse biomolecules. The subunit types of polymeric biomolecules differ based on their classification (Voet et al., 2008). A protein is a polymer composed of a minimum of 20 amino acids.

The field of biochemistry is concerned with studying the chemical reactions catalyzed by enzymes and analyzing the chemical properties of important biological molecules, such as proteins. Abundant information is accessible concerning the biochemistry of the endocrine system and cell metabolism. Additional areas of study in biochemistry include the genetic code (DNA, RNA), protein synthesis, cell membrane transport, and signal transduction (Voet et al., 2008).

Molecular biology and biochemistry have numerous similarities and their objectives are typically complementary. Biochemistry utilizes the techniques and findings of molecular biology to enhance our understanding of cellular processes such as cell signaling and apoptosis. On the other hand, molecular biology specifically concentrates on comprehending the molecular mechanisms involved in genetic material replication, transcription, and translation (Voet et al., 2008).

1.4.1. Transcription

DNA fulfills two functions in both prokaryotes and eukaryotes: firstly, it undergoes replication, and secondly, it encodes the instructions necessary for synthesizing the proteins essential for the cell to execute all of its functions. In order to achieve this, the DNA is transcribed into an mRNA molecule. Translation utilizes mRNA to provide the necessary genetic code for protein synthesis. The construction of a protein involves the utilization of an initial amino acid sequence, which is achieved through the processes of transcription and translation (Voet et al., 2008).

1.4.2. Interpretation

The synthesis of proteins is one of the most energy-intensive metabolic processes in a cell. Proteins, in contrast, perform a wide range of cellular functions and constitute the majority of all living organisms' components (excluding water). Translation, or protein synthesis, is the process by which an mRNA message is decoded to produce a polypeptide product. The covalent linkage between amino acids can result in polypeptide chains ranging in length from approximately 50 to over 1,000 amino acids.

1.4.3. Protein Structure Levels

Proteins are the primary functional components of cells. Proteins play a crucial role in virtually all cellular processes. Protein enzymes have a wide range of functions, such as facilitating communication, regulating the expression of genes, providing structure to cells and multicellular organisms, and catalyzing most biological processes. Proteins are essential for the existence of life as we know it. The adaptability of proteins can be attributed to the wide range of structures they possess (Voet et al., 2008).

1.4.4. Enzymes

The process of meat digestion through the action of stomach secretions and the conversion of starch into sugars by plant extracts and saliva was first acknowledged in the late 18th and early 19th centuries. However, the specific mechanism through which this occurred remained undisclosed. In the 19th century, Louis Pasteur identified a vital force known as "ferments" that facilitated the conversion of sugar into alcohol in yeast cells. Previously, it was widely held that fermentation could exclusively take place within living organisms. According to him, alcoholic fermentation is a process that is associated with the vitality and structure of yeast cells, rather than their decay or decomposition (Voet et al., 2008).

In 1878, German scientist Wilhelm Kühne (1837–1901) coined the term "enzyme," which means "in leaven," to describe this process. Subsequently, the term "ferment" was employed to denote the chemical activity generated by living organisms, whereas the term "enzyme" was utilized to describe inanimate molecules such as pepsin. In 1897, Eduard Buchner conducted research on the ability of yeast extracts to ferment sugar without the presence of live yeast cells. Through experiments conducted at the University of Berlin, it was found that the sugar continued to undergo fermentation even without the presence of live yeast cells. The enzyme responsible for sucrose fermentation was named "zymase" by him. In recognition of his biochemical research and his discovery of cell-free fermentation, he was awarded the Nobel Prize in Chemistry in 1907. Following Buchner's precedent, the name of an enzyme is determined by the reaction it catalyzes (Voet et al., 2008). The substrate, such as lactose, is cleaved by the enzyme lactase. Similarly, the enzyme DNA polymerase produces DNA. These enzymes are typically named with the suffix -ase.

1.5. METHODS EMPLOYED IN THE FIELD OF ANIMAL BIOCHEMISTRY

The core tenets of genetics encompass DNA replication, transcription, translation, and the central dogma of life. By employing basic molecular techniques, molecular genetics research has provided new opportunities for the identification of diseases. Recently, advanced molecular biology technology has become a valuable tool for discovering new genes that play a vital role in medicine, human health, animal breeding, agriculture, the environment, and other related fields.

Molecular genetics in biology investigates how variations in DNA molecule expression or structure contribute to the diversity of animals. In the field of molecular genetics, the process of examining genetic material is a commonly used method to explore the structure and/or function of an organism's genes (Ehtisham et al., 2016). Adenine, Guanine, Cytosine, and Uracil constitute the genetic code found in all organisms. By combining these base pairs and reading them in groups of three, specific amino acids are produced. The field of genetics is concerned with the transmission of genes across generations and their evolution over time. Contemporary genetic research combines with traditional genetics to understand molecular genetics, which examines an organism's macromolecules and analyzes DNA. The DNA or RNA derived from a specific gene is separated and analyzed using various molecular methods. This section discusses a limited selection of the numerous molecular methods employed in molecular genetics.

1.5.1. Isolating Genomic DNA

DNA, also known as deoxyribonucleic acid, is the hereditary material present in all organisms. In 1869, Friedrich Miescher successfully isolated DNA for the first time. DNA extraction is the procedure through which DNA is obtained from the cells of an organism. The DNA extraction process utilizes biological materials such as blood, saliva, or tissue. The extraction of DNA involves several procedures, such as cell disintegration, protein removal, and purification of the DNA to eliminate any remaining biological components. Subsequently, purified DNA is employed for various procedures such as cloning, sequencing, or polymerase chain reaction (Ehtisham et al., 2016).

The three predominant methods for DNA extraction include mechanical, enzymatic lysis, and chemical processes. The process of isolating DNA commonly involves three steps: precipitation, filtration, and concentration. The techniques most commonly used are alcohol precipitation, phenol-chloroform extraction, and silica-based purification. There is currently a large selection of DNA extraction kits available for purchase to meet the demand for genomic DNA kits (Ehtisham et al., 2016).

DNA isolation involves the processes of cell lysis, precipitation, purification, and concentration.

1. **Cell Lysis:** Cell lysis refers to the procedure of rupturing cells in order to liberate DNA. For example, the combination of a solution salt

such as SDS with a detergent is commonly employed to disintegrate bacterial cell membranes and liberate DNA. The enzymatic and mechanical methods of DNA extraction are commonly used in both plant and animal cells.

2. **DNA Precipitation:**Once the DNA of the cell has been released, remove any remaining components, such as pollutants and cell proteins. In this phase, precipitating agents such as alcohol (ethanol or isopropanol) or salt (ammonium acetate) are commonly employed. By implementing this methodology, we ascertained the presence of residual impurities in the liquid and a solid mass of DNA settled at the bottom of the solution.
3. **DNA Purification:**Column-based methods are employed to enhance the purification of genetic material after DNA precipitation. Illustrations of this nature encompass silica-based spin columns that have an affinity for DNA binding, the employment of centrifugation to eliminate impurities from cells, and the utilization of centrifugal force to precipitate DNA at the bottom of a tube, thereby facilitating purification.
4. **DNA Concentration:**The DNA is concentrated by removing the excess liquid from the tube. Two commonly employed techniques for preservation are freeze drying and vacuum centrifugation. The requirement for effective DNA extraction procedures in medical microbiology has been emphasized by the demand for molecular diagnostic assays. Moreover, the preference is for the automation of these techniques. There are multiple DNA isolation kits available in the market that facilitate the quick and efficient separation of DNA from various materials. A standard DNA isolation kit follows a series of steps: alkaline lysis, incubation with a DNA binding matrix, DNA washing, and finally elution. Some commercially available brands of DNA isolation kits include Thermofisher, Sigma-Aldrich, Qiagen, and BioRad (Ehtisham et al., 2016).

1.5.2. Polymerase Chain Reaction (PCR)

The polymerase chain reaction (PCR) is a molecular technique that utilizes enzymes to exponentially synthesize large quantities of a specific DNA target region *in vitro*. The PCR technique for DNA synthesis employs the DNA polymerase enzyme and closely resembles the process of DNA synthesis that occurs naturally within living cells. Kary Mullis adopted the use of PCR during the mid-1980s. PCR is a contemporary molecular technique that assists in the characterization and identification of various species, such as fungi, viruses, bacteria, and parasites (Ehtisham et al., 2016).

1.5.3. Restriction Digestion and DNA Ligation

Restriction enzymes, which are utilized in restriction digestion, hydrolyze DNA by cleaving its phosphodiester bonds. Every enzyme cleaves the double stranded DNA at or near the target sequences after identifying them. Additionally, they serve as a constituent of the defense mechanisms that bacteria utilize to protect themselves against viruses and other forms of bacterial DNA. Various restriction enzymes generate single-stranded DNA overhangs by creating staggered cuts. Conversely, some paths result in no progress or success. DNA ligase is the most commonly employed enzyme for joining DNA strands (Ehtisham et al., 2016).

Ligase is employed in the mending of fragmented DNA molecules in cases where two DNA strands possess indistinguishable terminations.

1.6. APPLICATIONS OF ANIMAL BIOCHEMISTRY

Comprehending the biochemistry of animals is crucial for understanding their interactions with the environment (Kaneko et al., 2008). The information provided includes details regarding the animal's metabolism, dietary habits, general well-being, and level of productivity. This information is essential for occupations such as farming, veterinary medicine, and the pet industry. The field of animal biochemistry significantly influences agricultural and animal productivity through various means.

1. **Nutrition Optimization:** Developing a well-balanced diet for cattle is facilitated by a comprehension of animal biochemistry. Optimal nutrition improves overall health, reproductive capacity, and growth rates.
2. **Disease Prevention:** Our comprehension of biochemistry enables us to develop vaccinations and medications for the prevention and treatment of cattle diseases. Healthy animals exhibit higher productivity.
3. **Genetics and Breeding:** The process of selectively breeding organisms to obtain specific traits is influenced by the principles of biochemistry. Genetic engineering enhances both the quantity and quality of animals.
4. **Efficiency and Metabolism:** Researching metabolic pathways enhances feed conversion efficiency. Efficient metabolism reduces resource wastage.

5. **Environmental Impact:** The application of biochemistry, as demonstrated by Kaneko et al. (2008), provides valuable insights into sustainable methods for mitigating pollution caused by animal agriculture. To enhance agricultural productivity and sustainability, farmers can proactively incorporate the following biochemical principles into their daily practices:
6. **Nutrient Management:** Understand the specific biochemical needs of plants for different nutrients.
   - Utilize fertilizers that are carefully formulated to optimize crop growth.
   - Utilize biochemistry to develop non-toxic insecticides for the purpose of managing diseases and pests.
   - Plant health can be monitored using biochemical indicators.
7. **Crop Breeding:** Employ biochemistry to selectively determine desired characteristics in crops. Develop disease-resistant cultivars.
8. **Soil Health:** Employ biochemical assays to examine the constitution of the soil.
9. **Livestock Nutrition:** Apply principles of biochemistry to understand animal metabolism. Develop nutrition plans that ensure a harmonious balance of nutrients for livestock.

Ultimately, the selection of sustainable farming methods and strategies to enhance crop productivity is guided by the principles of biochemistry (Kaneko et al., 2008).

1.7. THE FUTURE OF ANIMAL BIOCHEMISTRY

The field of animal biochemistry has undergone substantial expansion and progress in recent years, propelled by scientific advancements, globalization, and heightened social awareness. The worldwide biomedical research industry has grown rapidly, resulting in a rise in the utilization of animals for testing, teaching, and research on a global scale. The rate of scientific output has experienced exponential growth, with the quantity of published information doubling every nine years. The result of this has been an increased consciousness among the general public and decision-makers regarding the significance of animal welfare and ethical deliberations in biomedical research (Bert et al., 2016).

To address these concerns, there has been an increased frequency in the implementation and updating of laws, rules, and guidelines regarding the welfare of laboratory animals. The purpose of these measures is to tackle public concerns, enhance the health and well-being of research animals, and establish uniform procedures in facility engineering, animal care, veterinary services, and program monitoring (Bert et al., 2016). Nevertheless, there is a continuous dispute regarding the sufficiency and superfluity of the supervision and regulations pertaining to the utilization of animals in scientific investigations. There is a debate between those who support more stringent regulations and those who believe that the scientific community should regulate itself (Bert et al., 2016).

A significant obstacle encountered by the biomedical research community is the onerous regulatory requirements linked to animal research (Bert et al., 2016). The regulatory compliance procedures are frequently perceived as challenging and onerous, resulting in an emphasis on superfluous administrative protocols that do not enhance animal welfare or research excellence. The governance and reporting obligations enforced by various federal agencies also present difficulties for research institutions, amplifying financial burdens and the volume of documentation necessary. The challenges are worsened by the decrease in funding for biomedical research (Bert et al., 2016).

Aside from regulatory obstacles, the animal research community is also confronted with emerging legal challenges that affect animal care and usage programs. Activist groups have been filing a growing number of legal challenges concerning issues such as public disclosure, animal transportation, guardianship, and the legal rights of animals. These challenges diverge from the ethical rationales for utilizing animals in research and present a danger to animal research.

The transportation of research animals is a matter of concern for the animal research community. Several commercial airlines have ceased transporting animals used for research as a result of the influence exerted by animal rights organizations. This gives rise to logistical complexities and escalates the expenses associated with animal experimentation, which could potentially impede or bring to a halt research endeavors. Although certain countries provide government aid for the transportation of research animals, there are no regulations mandating airlines to transport animals specifically for research objectives. The absence of regulation and the freedom of commercial airlines to decide whether or not to transport animals for research purposes make the transportation of research animals more complex (Bert et al., 2016).

In order to tackle these difficulties, it is imperative to foster collaboration among research institutions, scientists, veterinarians, managers, and other personnel involved in supporting animal research. Establishing consistent regulations and reporting standards, while also promoting the exchange of information among research organizations, government authorities, oversight committees, and accreditation bodies, can simplify the regulatory landscape for animal research. Effective communication of the advantages of animal research to the public is crucial for the scientific community and funding organizations. This involves emphasizing the positive impact on both human and animal health that has been achieved through such research (Bert et al., 2016).

Ultimately, the domain of animal biochemistry has experienced substantial progress and development in recent times. Nevertheless, the growing utilization of animals in scientific research has generated apprehensions regarding animal well-being and ethical deliberations. The animal research community faces additional challenges due to the regulatory burden and emerging legal challenges associated with their work. To tackle these challenges and ensure the ongoing advancement of animal biochemistry research, it is imperative to engage in collaboration, harmonize regulations, and enhance communication with the public (Bert et al., 2016).

Chapter 2

Animal Carbohydrate Metabolism

Contents

Carbohydrate metabolism refers to the complete set of molecular reactions involved in the synthesis, breakdown, and utilization of carbohydrates in living organisms. Carbohydrates play a crucial role in numerous essential metabolic pathways. Plants utilize photosynthesis to convert carbon dioxide and water into carbohydrates, allowing them to store solar energy internally. Animals and fungi utilize cellular respiration to metabolize the stored carbohydrates in plants, enabling cells to access and utilize the energy. Plants and animals temporarily store the energy they release in high-energy molecules such as adenosine triphosphate (ATP), which is used for various cellular functions (BeMiller, 2018).

2.1. OVERVIEW OF CARBOHYDRATE STRUCTURE AND FUNCTION

Carbohydrates are biomolecules that contain multiple hydroxyl groups and belong to a class of naturally occurring carbonyl compounds, such as aldehydes or ketones. The substance consists of carbon (C), hydrogen (H), and oxygen (O) atoms, usually in a hydrogen-to-oxygen ratio of 2:1, similar to that of water. The empirical expression for it is Cm (H_2O) n or (CH_2O) n, with the possibility of m and n being different or not. Nevertheless, certain substances, such as uronic acids, deviate from this precise stoichiometric description. In addition, certain substances such as formaldehyde and acetic acid, although they have similar chemical structures to carbohydrates, are not classified as carbohydrates (BeMiller, 2018).

Carbohydrates are categorized into four primary groups according to their polymerization level: monosaccharides, disaccharides, oligosaccharides, and polysaccharides. Below is a concise description of the composition and roles of carbohydrate groups.

2.1.1. Monosaccharides

Monosaccharides are the most basic form of carbohydrates because they cannot be broken down into smaller units through hydrolysis. The term "mono" in monosaccharides (Figure 2.1) signifies the presence of a single sugar unit. They function as the fundamental building blocks for polysaccharides and disaccharides. As a consequence, they are also known as simple sugars. Simple sugars are white, crystalline solids that are soluble in water but insoluble in nonpolar solvents. The general formula for the monosaccharide structure is (H_2O) n, or $C_nH_{2n}O_n$. The smallest monosaccharides consist of dihydroxyacetone and D- and L-glyceraldehydes, with a value of n equal to 3 (BeMiller, 2018).

Ketoses are monosaccharides that contain ketone groups, which have the structure RC(=O)R′. On the other hand, aldolases are monosaccharides that have aldehyde groups, which have the structure R-CHO. Some examples of monosaccharides are glucose, fructose, ribulose, and erythrulose. D-glucose is the most common, abundant, and widely distributed carbohydrate. Dextrose, also known as aldohexose, is an aldehyde compound consisting of six carbon atoms. It is present in both cyclic and open-chain configurations. The majority of monosaccharides have names that end in -ose. These compounds can also be called trioses (containing three carbons), tetroses (containing four carbons), pentoses (containing five carbons), hexoses (containing six carbons), and heptoses (containing seven carbons), depending on the number of carbons present. Typically, the number of carbons in these compounds ranges from three to seven (BeMiller, 2018).

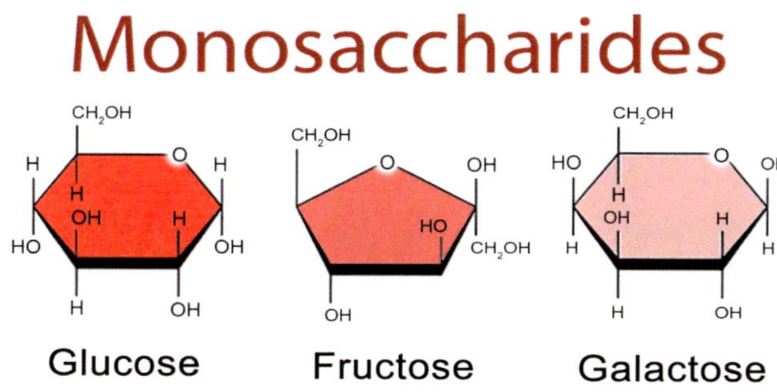

Figure 2.1. *Examples of monosaccharides.*

Source: Modric (2016).

2.1.1.1. Functions of Monosaccharides

- Both plants and humans depend on glucose ($C_6H_{12}O_6$) as a primary source of energy. Plants utilize carbon dioxide and water to synthesize glucose, which fulfills their energy requirements. The surplus glucose is transformed into starch, which serves as a source of nutrition for humans and other herbivores.
- Lactose, a type of sugar found in milk, consists of galactose. On the other hand, fructose, present in fruits and honey, is responsible for the sweet taste of these items.
- Ribose is a fundamental building block of various coenzymes and nucleic acids.

- Mannose is a crucial constituent of glycoproteins and mucoproteins, serving essential functional roles.

2.1.2. Disaccharides

Disaccharides consist of two sugar units (Widmalm, 2013) (Figure 2.2). During a dehydration reaction, also referred to as a condensation reaction or dehydration synthesis, two monosaccharide units are liberated.

During this process, the hydroxyl group of one monosaccharide forms a covalent bond with the hydrogen of another monosaccharide, resulting in the release of a water molecule. A glycosidic bond refers to the covalent linkage that occurs between two sugar molecules. Possible glycosidic linkages or bonds include alpha or beta types. When the hydroxyl (OH) group attached to the carbon-1 of the first glucose molecule is positioned below the plane of the ring, it forms an alpha bond. Conversely, when the OH group is positioned above the ring plane, it forms a beta bond (Widmalm, 2013).

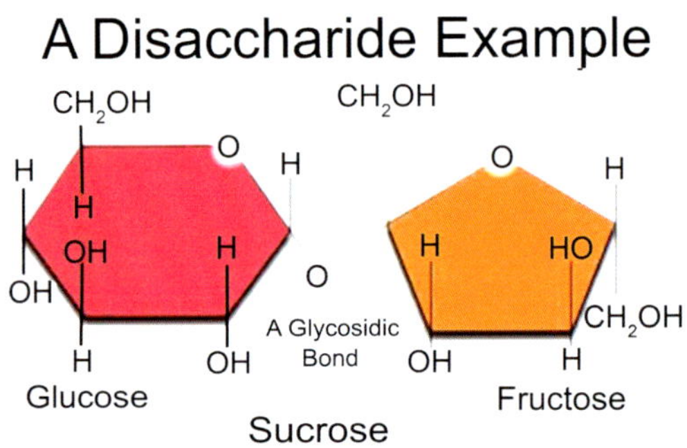

Figure 2.2. *A disaccharide molecule is formed by 2 monosaccharides, joined by a glycosidic bond.*

Source: Modric (2015).

Lactose, maltose, and sucrose are three examples of disaccharides. Sucrose, the most abundant disaccharide, is composed of a single D-glucose molecule and a single D-fructose molecule. The systematic name for sucrose is O-α-D-glucopyranosyl-(1→2)-D-fructofuranoside.

Lactose is a naturally-occurring disaccharide composed of one molecule of D-galactose and one molecule of D-glucose. It is commonly found in the milk of mammals. Lactose is scientifically referred to as O-β-D-galactopyranosyl-(1–4)-D-glucopyranose.

Disaccharides can be classified into two categories based on their ability to undergo oxidation-reduction reactions.

- **Sugar Reduction:** A disaccharide that acts as a reducing aldehyde group through a free hemiacetal unit. Cellobiose and maltose are two instances.
- **Non-Reducing Sugar:** Disaccharides that are connected by an acetal linkage between their anomeric centers, resulting in the absence of a free hemiacetal group. Trehalose and sucrose are two instances.

2.1.2.1. The Function of Disaccharides

- Photosynthesis, which is a primary source of carbon and energy for plants, produces sucrose as a byproduct.
- Animals primarily derive their energy from lactose.
- Maltose is a crucial component in the breakdown of starch and glycogen during digestion.
- Trehalose is an essential energy source for insects.
- Cellobiose is necessary for the metabolism of carbohydrate.
- Gentiobiose is found in plant glycosides and specific polysaccharides.

2.1.3. Oligosaccharides

Oligosaccharides are compounds that, when broken down by hydrolysis, yield 3 to 10 molecules of either unique or identical monosaccharides. The glycosyl linkage connects all the monosaccharides together. In addition, oligosaccharides are classified as trisaccharides, tetra saccharides, pent saccharides, and so forth, based on the number of monosaccharides they are composed of (Widmalm, 2013). Figure 2.3 depicts the examples of oligosaccharides.

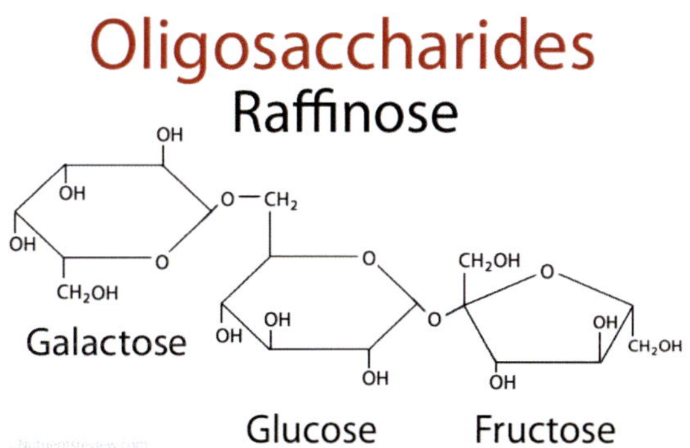

Figure 2.3.*Examples of oligosaccharides.*

Source: Modric (2017).

Trisaccharides follow the general formula Cn (H_2O) n–2, while tetra saccharides adhere to the formula Cn (H_2O) n–3, and so forth. Typically, oligosaccharides are present in the form of glycans. These compounds, known as glycolipids or glycoproteins, are linked to lipids or amino acid side chains in proteins through N- or O-glycosidic linkages.

Glycosidic bonds form through glycosylation, a process that involves the covalent bonding of a carbohydrate to an organic molecule. This process results in the formation of structures such as glycolipids and glycoproteins (Widmalm, 2013).

N-Linked Oligosaccharides: These oligosaccharides are attached to asparagine by forming a beta linkage with the amine nitrogen in the side chain. Eukaryotes carry out this process at the membrane of the endoplasmic reticulum. Conversely, it occurs at the plasma membrane in prokaryotes.

O-Linked oligosaccharides refer to oligosaccharides that are connected to the hydroxyl group of serine or threonine on the side chain. The Golgi apparatus is responsible for the addition of monosaccharide units to a complete polypeptide chain.

2.1.3.1.The Mechanisms and Functions of Oligosaccharides

- Carbohydrates and proteins combine to create glycoproteins, which have important functions in terms of antigenicity, solubility, and protease resistance. Glycoproteins play a crucial role as tumor antigens, cell-surface receptors, immunoglobulins, and molecules that enhance cell adhesion.
- Glycolipids, which are lipids with attached carbohydrates, impact membrane proteins that serve as receptors and play a vital role in cell recognition.
- Cell adhesion through oligosaccharides is facilitated by lectins, which are specific protein complexes that attach to carbohydrates (Widmalm, 2013).

2.1.4. Polysaccharides

A polysaccharide is a chain of over 10 carbohydrates that are linked together through the formation of glycosidic bonds. They have a broad distribution and primarily serve an organism's structural or storage requirements. Glycans, also known as glycans, are another term used to refer to them.

The composition and structure of the molecules that bind or react with these substances, along with their interactions with the enzyme machinery, dictate their physical and biological characteristics. Polysaccharides are categorized based on their origin, the type of monosaccharide units they contain, or their functions.

Polysaccharides are classified into two categories: homopolysaccharides and heteropolysaccharides, based on the type of monosaccharides used to form their structures.

2.1.4.1. The Function of Polysaccharides

Structural polysaccharides (Figure 2.4) provide mechanical stability to organisms, organs, and cells. Cellulose and chitin are two examples. Cellulose is an essential component of the diet of ruminant animals, while chitin plays a role in the construction of fungal cell walls.

Polysaccharides for storage serve as reservoirs of carbohydrates that can be broken down into individual sugar molecules when the body requires them. Examples of polysaccharides include glycogen, inulin, and starch. Plants utilize starch as an energy storage, while animals rely on it to fulfill their energy requirements via the activity of the salivary enzyme amylase. Animals, bacteria, and fungi utilize glycogen as a polysaccharide for energy storage, while plants employ inulin for the same purpose.

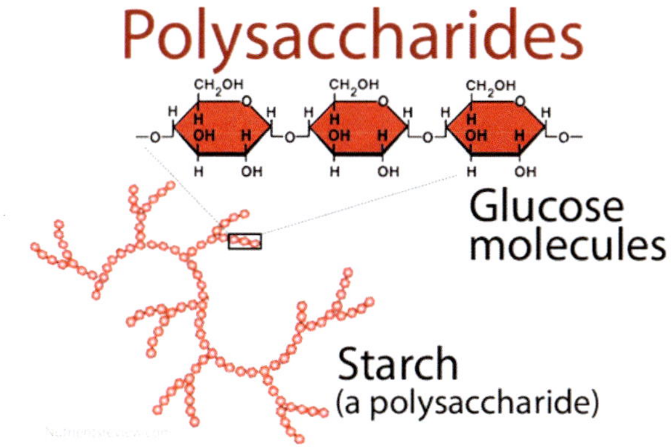

Figure 2.4. *Polysaccharides structure.*

Source: Modric (2016).

2.2. GLYCOLYSIS IN ANIMALS

Glycolysis is a sequence of chemical reactions that convert glucose into either lactate or pyruvate. Regarded as a paradigm for all metabolic pathways, this pathway holds the distinction of being the first one to be comprehensively understood (Rogatzki et al., 2015). The Embden-Meyerhoff pathway is an alternative term for glycolysis. The cytoplasm of nearly all animal cells serves as the site for the complete sequence of reactions. Oxygen molecules are absent throughout the entire process. Glycolysis yields a net gain of four ATP molecules while utilizing two ATP molecules. Consequently, the procedure yields two net ATP molecules. In addition, two molecules of NADH are generated during the process. The destiny of pyruvate is influenced by the presence or absence of oxygen in cells containing mitochondria or by the presence or absence of mitochondria in the cell (Rogatzki et al., 2015).

Glycolysis begins with a six-carbon ring structure of a glucose molecule and ends with the production of two molecules of the three-carbon sugar pyruvate. Glycolysis occurs in two distinct phases. To ensure the equal division of the six-carbon sugar molecule into two three-carbon molecules, the initial step of the glycolysis pathway involves keeping the glucose molecule within the cell and expending energy to modify it. During the second stage of glycolysis, energy is derived from the molecules and subsequently stored as ATP and NADH. NADH refers to NAD in its reduced state (Rogatzki et al., 2015).

** Please include TCA cycle with energectics and physiological importance which is a important part of of glycolysis. May elaborate aerobic and anaerobic glycolysis.

2.2.1. Initial Phase of Glycolysis Involving Energy-Consuming Steps

- **Step 1:** Hexokinase, a versatile enzyme, facilitates the addition of a phosphate group to six-carbon sugars, initiating the process of glycolysis. Hexokinase utilizes ATP as the phosphate source to catalyze the phosphorylation of glucose, resulting in the formation of glucose-6-phosphate, a more reactive form of glucose. Due to phosphorylation, the glucose molecule is incapable of binding with the GLUT proteins. Additionally, it cannot leave the cell because the negatively charged phosphate hinders its passage through the hydrophobic interior of the plasma membrane (Rogatzki et al., 2015).
- **Step 2:** During the second step of glycolysis, an isomerase enzymatically converts glucose-6-phosphate into fructose-6-phosphate, which is an isomer of glucose-6-phosphate. An isomerase is an enzyme that facilitates the conversion of a molecule into one of its isomers. The sugar can undergo a conversion from phosphoglucose to phosphofructose, resulting in the splitting of the sugar molecule into two molecules, each containing three carbons.
- **Step 3:** The phosphofructokinase enzyme facilitates the process

of phosphorylating fructose-6-phosphate. Fructose-6-phosphate is phosphorylated by a second ATP molecule, resulting in the formation of fructose-1,6-bisphosphate. Phosphofructokinase is an enzyme that regulates the speed of this process. When the concentrations of ADP are elevated, it becomes active; however, when the concentrations of ATP are elevated and the levels of ADP are low, it becomes less active. Consequently, the route decelerates when there is a "adequate" amount of ATP in the system. ATP, being the ultimate outcome of glucose catabolism, exhibits end product inhibition (Rogatzki et al., 2015).

- **Step 4:** The introduction of high-energy phosphates further destabilizes fructose-1,6-bisphosphate. The enzyme aldolase produces dihydroxyacetone-phosphate and glyceraldehyde-3-phosphate, which are two three-carbon isomers of 1,6-bisphosphate, in the fourth stage of glycolysis.
- **Step 5:** In the fifth step, an isomerase enzyme transforms dihydroxyacetone-phosphate into its isomer, glyceraldehyde-3-phosphate. Therefore, two molecules of the identical isomer will progress along the pathway. During this phase of the process, a single glucose molecule is metabolized with a net energy expenditure of two ATP molecules. Figure 2.5 discusses the glycolytic pathway in animals.

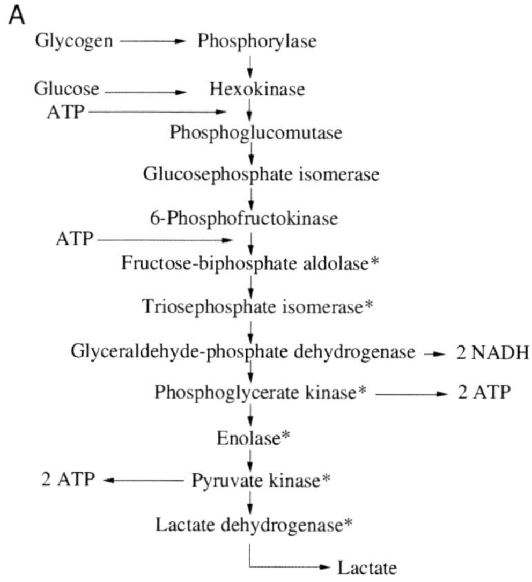

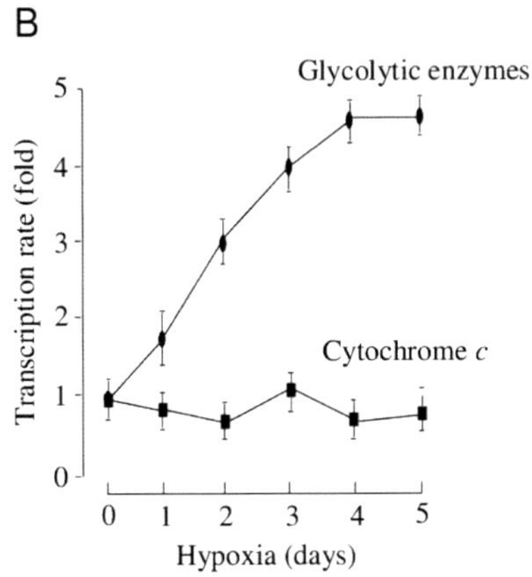

Figure 2.5. *The input discusses the glycolytic pathway in animals, including its composition, regulation, and induction by hypoxia.*

Source: Webster (2003).

2.2.2. Energy-Releasing Steps of the Second Half of Glycolysis

During glycolysis, the cell has utilized two ATP molecules and produced two small, three-carbon sugar molecules. Both of these molecules will proceed through the second phase of the pathway, where sufficient energy will be extracted to recover the initial expenditure of two ATP molecules and produce two additional ATP molecules and two NADH molecules with even higher energy levels for the cell.

- Step 6: In the sixth step of glycolysis, the sugar glyceraldehyde-3-phosphate undergoes oxidation, resulting in the release of high-energy electrons. These electrons are subsequently captured by the electron carrier NAD^+ to generate NADH. Subsequently, a second phosphate group is appended to the sugar, resulting in its phosphorylation and the formation of 1,3-bisphosphoglycerate. It is important to mention that the addition of the second phosphate group does not require any extra ATP molecule.
- Step 7: During the seventh step, 1,3-bisphosphoglycerate transfers a phosphate group with high energy to ADP, resulting in the formation of one ATP molecule. This process is facilitated by an enzyme called phosphoglycerate kinase, which is named after the reverse reaction. This is a depiction of phosphorylation occurring at the substrate level. 3-phosphoglycerate is formed through the oxidation

of the carbonyl group on 1,3-bisphosphoglycerate, resulting in the formation of a carboxyl group.

- •**Step 8:** The conversion of 3-phosphoglycerate to 2-phosphoglycerate involves the relocation of the phosphate group from the third carbon to the second carbon, resulting in the formation of an isomer. The occurrence of this phase is facilitated by an enzyme known as mutase (isomerase).
- **Step 9:** The ninth step is facilitated by the enzyme enolase. This enzyme catalyzes a dehydration reaction in 2-phosphoglycerate, leading to the formation of a double bond. This double bond increases the potential energy in the remaining phosphate bond and produces phosphoenolpyruvate (PEP).
- **Step 10:** Pyruvate kinase, an enzyme named after the reverse reaction of converting pyruvate to PEP, catalyzes the last step in glycolysis. This step produces pyruvic acid (or pyruvate in its salt form) and an additional ATP molecule through substrate-level phosphorylation. Due to their ability to facilitate both forward and reverse reactions, numerous enzymes in enzymatic pathways are named after the reverse reaction. This reverse reaction is often defined by the reaction that occurs *in vitro*, under non-physiological conditions.

2.2.3. Outcomes of Glycolysis

Glycolysis starts with glucose and produces two pyruvate molecules, four ATP molecules, and two NADH molecules at the end. The cell currently possesses a total increase of two ATP molecules and two NADH molecules available for utilization. This increase is a result of the consumption of two ATP molecules in the initial phase of the pathway, which was necessary to prepare the six-carbon ring for cleavage. If the cell is incapable of further catabolizing the pyruvate molecules, it will only be able to extract two ATP molecules from a single glucose molecule. Mature mammalian red blood cells exclusively generate ATP *via* glycolysis due to their inability to engage in aerobic respiration, the metabolic process through which organisms convert energy in the presence of oxygen. If glycolysis is halted, these cells lose the ability to sustain their sodium-potassium pumps, leading to their eventual demise (Rogatzki et al., 2015).

Insufficient pyruvate kinase will prevent the final step of glycolysis, which involves the synthesis of pyruvate, from occurring. The entire glycolysis pathway will function in this scenario, however, only two ATP molecules will be generated in the latter half. Pyruvate kinase is an enzyme that regulates the speed of glycolysis.

2.3. GLUCONEOGENESIS IN ANIMALS

** Please include pathway figure of gluconeogenesis

Gluconeogenesis (GNG) is a metabolic process that synthesizes glucose from non-carbohydrate carbon sources. It is a ubiquitous biological process observed in all living organisms, including bacteria, fungi, plants, and other microorganisms. Gluconeogenesis primarily occurs in the liver and, to a lesser extent, in the kidney cortex of vertebrates. It is a primary method used by humans and other animals to control blood sugar levels and avoid hypoglycemia, with the alternative being the breakdown of glycogen known as glycogenolysis. Gluconeogenesis occurs in ruminants regardless of factors such as fasting, low-carb diets, exercise, etc., due to the ability of rumen organisms to digest dietary carbohydrates. The process occurs in various other animals during periods of fasting, starvation, low-carb diets, or intense physical activity.

According to Miyamoto & Amrein (2017), humans can use any non-carbohydrate source that can be converted into pyruvate or glycolysis intermediates as substrates for gluconeogenesis. The substrates involved in these processes consist of glucogenic amino acids (excluding ketogenic amino acids) for protein breakdown, glycerol and odd-chain fatty acids (excluding even-chain fatty acids) for lipid breakdown (such as triglycerides), and lactate from the Cori cycle for other metabolic pathways. Acetone derived from ketone bodies can serve as a substrate during prolonged fasting, creating a pathway for fatty acids to be converted into glucose. Both long-term fasting and diabetes increase the kidney's relative involvement in gluconeogenesis, despite the fact that the liver is responsible for the majority of it (Miyamoto & Amrein, 2017).

The gluconeogenesis pathway is primarily endergonic, meaning it requires energy input from the hydrolysis of ATP or GTP. As a result, the overall process becomes exergonic. To illustrate, the spontaneous progression of the route from pyruvate to glucose-6-phosphate requires four molecules of ATP and two molecules of GTP. The process of beta oxidation is responsible for the breakdown of fatty acids, resulting in the production of ATP molecules.

2.3.1. Precursors

Propionate is the primary substance used by ruminants for the production of glucose. Propionate is synthesized through the β-oxidation process of odd- and branched-chain fatty acids in nonruminant animals, including humans. It serves as a minor substrate for gluconeogenesis.

In the liver, lactate is metabolized through the Cori cycle via the enzyme lactate dehydrogenase, resulting in the production of pyruvate. Subsequently, pyruvate, which is the initial substrate of the gluconeogenic pathway, can be utilized to generate glucose (Miyamoto & Amrein, 2017). The carbon framework of amino acids can be incorporated into the cycle either directly (as pyruvate or oxaloacetate) or indirectly (through the citric acid cycle) due to the processes of transamination and deamination. The duration of fasting directly impacts the proportion of glucose synthesis that is attributed to lactate produced by the Cori cycle.

According to Miyamoto & Amrein (2017), during fasting periods of 12, 20, and 40 hours, the contribution of lactate from the Cori cycle to gluconeogenesis was found to be 41%, 71%, and 92% respectively, as observed in human volunteers.

An enduring question in the field of biology is whether animals have the ability to convert fatty acids with an even number of carbon atoms into glucose. Odd-chain fatty acids can be oxidatively converted to generate propionyl- and acetyl-CoA. Propionyl-CoA serves as a precursor to succinyl-CoA, which can subsequently be converted into oxaloacetate to initiate the gluconeogenic process. On the other hand, acetyl-CoA is the only result of the breakdown of fatty acids with an even number of carbon atoms. To use acetyl-CoA in the process of gluconeogenesis, a glyoxylate cycle, also known as a glyoxylate shunt, is required. This cycle produces precursors for four-carbon dicarboxylic acids (Miyamoto & Amrein, 2017). Bacteria, plants, and fungi share the presence of two enzymes, isocitrate lyase and malate synthase, which together constitute the glyoxylate shunt.

The genes responsible for both enzymatic functions have only been identified in worms, where they are found as a single enzyme capable of performing both functions. However, there have been some reports of glyoxylate shunt enzymatic activities in mammalian tissues. Additional organisms, including arthropods, echinoderms, and specific vertebrates, have demonstrated the presence of genes encoding only malate synthase, without isocitrate lyase. Placental mammals lack the malate synthase gene, whereas monotremes (such as the platypus) and marsupials (like the opossum) possess it.

2.3.2. Pathway

Gluconeogenesis is a metabolic pathway that involves 11 enzymatic reactions. The route will commence either in the cytoplasm or mitochondria of the liver or kidney, depending on the substrate being utilized. There are multiple reactions that are contrary to the steps of glycolysis.

- Pyruvate undergoes carboxylation in the mitochondria to produce oxaloacetate, which initiates gluconeogenesis. Pyruvate carboxylase serves as the catalyst for this process, requiring one ATP molecule.

The liver produces high levels of acetyl-CoA during β-oxidation, which in turn activates this enzyme. Conversely, high levels of ADP and glucose inhibit its activity.

- NADH is required for the conversion of oxaloacetate to malate, which is subsequently transported out of the mitochondria.
- The cytosol is where the conversion of malate to oxaloacetate takes place during gluconeogenesis, using NAD^+ as a catalyst.
- The enzyme PEPCK removes a carboxyl group from oxaloacetate and adds a phosphate group to produce phosphoenolpyruvate. During this reaction, a process called hydrolysis breaks down a molecule of GTP into GDP.
- The reaction follows the same steps as the reverse of glycolysis. Fructose 1,6-bisphosphatase utilizes a single water molecule and releases one phosphate group to convert fructose 1,6-bisphosphate into fructose 6-phosphate. In contrast, phosphofructokinase 1 in glycolysis converts F6P and ATP into F1,6BP and ADP. This stage is also known as the gluconeogenic stage, in which the rate is restricted (Miyamoto & Amrein, 2017).
- Phosphoglucoisomerase catalyzes the conversion of fructose 6-phosphate to glucose-6-phosphate, which is the reverse of the second step in glycolysis. Glucose-6-phosphate can be utilized by various metabolic pathways, or it can undergo dephosphorylation to liberate glucose. Glucose-6-phosphate, which is the phosphorylated version of glucose, is confined within the cell and serves as the mechanism through which cells control their internal glucose levels. Conversely, unbound glucose can easily enter and exit the cell.
- Glucose-6-phosphate is broken down by glucose-6-phosphatase in the endoplasmic reticulum, resulting in the production of glucose and the release of an inorganic phosphate. This is the last step in the gluconeogenesis process, which leads to the creation of glucose. Hexokinase facilitates the transformation of glucose and ATP into G6P and ADP in the process of glycolysis. However, unlike the preceding two steps, this particular step does not merely involve the straightforward reversal of glycolysis. Glucose transporters facilitate the movement of glucose into the cytoplasm (Miyamoto & Amrein, 2017).

The pathway of gluconeogenesis involves crucial molecules and enzymes. The steps in glycolysis have opposite counterparts in many other processes.

2.4. GLYCOGEN METABOLISM IN ANIMALS

Animals have developed a mechanism for storing glucose that reduces the need for breaking down proteins to obtain substances needed for glucose production in the meantime. Glycogen serves as the primary means by which animals store glucose. Storing glucose as a polymer reduces the osmotic load inside cells, resulting in less water needed for hydration and an increased energy density of stored glucose (Gardner et al., 2014). Glycogen is present in granulated form, known as β-particles, within the cytosol and endoplasmic reticulum of muscle cells. Each β-particle consists of a single glycogen molecule. The β-particles in the liver undergo fusion to produce larger, rosette-shaped α-particles that can be observed using an electron microscope. Glycogen typically consists of several million molecules, each containing an average of 10,000 to 50,000 glucose residues. The enzymes necessary for the production and degradation of glycogen, as well as the regulation of these processes, are also found in the storage granules (Gardner et al., 2014).

Glycogen is present in nearly all cells of the body, with the liver and skeletal muscle having the highest concentrations. The amount of glycogen stored in tissues can vary significantly depending on metabolic and physiological needs. However, in a person at rest after a meal, the liver typically holds 4–7% of its wet weight in glycogen, while the muscle contains approximately 1%. The total quantity of glycogen stored in muscle is greater than that in the liver due to the fact that muscle tissue constitutes 10 times the body weight of hepatic tissue.

During periods of intense muscle contraction, when there is a lack of sufficient oxygen and energy in the blood, the muscle needs a quick and immediate source of glucose to support anaerobic glycolysis. The liver primarily oxidizes fatty acids, but it also regulates blood glucose levels during short periods of fasting and adjusts the availability of energy sources to meet the metabolic demands of different tissues in different physiological conditions. The level of liver glycogen is primarily influenced by the availability of glucose and gluconeogenic precursors. Muscle glycogen reserves are influenced by the rate of oxidative metabolism (tissue respiration) and muscular contraction. However, they are not strongly affected by dietary cues (Gardner et al., 2014).

Phosphoglucomutase is the sole enzyme that connects the separate metabolic pathways of glycogenolysis (the breakdown of glycogen) and glycogenesis (the synthesis and storage of glycogen). The regulation of glycogen production and breakdown often occurs in a reciprocal manner, where the promotion of one process inhibits the other. The control mechanisms for glycogen pathways are less interconnected compared to glycolysis and gluconeogenesis due to the fact that glycogen pathways ensure the availability of a single substrate, while glycolysis and gluconeogenesis produce and metabolize intermediates from multiple metabolic pathways.

2.4.1. Glycogen Synthesis

** Please include pathway figure of glycogen synthesis

The initial stage of glycogenesis involves the enzymatic addition of a phosphate group to glucose, catalyzed by glucokinase in the liver and hexokinase in the muscles and other tissues. During the second stage of glycogenesis, phosphoglucomutase facilitates a process that is analogous to that of phosphoglyceromutase, transforming glucose-6-phosphate into glucose-1-phosphate. In the third stage, glucose-1-phosphate is converted into uridine diphosphate (UDP) glucose by reacting with uridine triphosphate (UTP). UDP is the direct precursor of glycogen synthesis. UDP-glucose pyrophosphorylase, also referred to as glucose-1-phosphate uridylyl transferase, is the specific enzyme responsible for facilitating this biochemical reaction. The reaction exhibits reversible characteristics, however, the action of pyrophosphatase results in the hydrolysis of pyrophosphate (PPi), rendering it almost irreversible (Gardner et al., 2014).

$$PPi + H_2O \rightarrow 2Pi$$

To produce UDP-glucose, it is necessary to break down two high-energy phosphate bonds through hydrolysis. Using nucleoside diphosphate sugars as intermediaries in carbohydrate condensation processes is a widely accepted practice. The energy needed to link the new sugar to the non-reducing portion of another sugar or polysaccharide is provided by the high-energy bond between the sugar and the nucleoside diphosphate.

The next step in the formation of glycogen involves the addition of a glucosyl residue at the C-1 position of UDP-glucose to a tyrosyl group at position 194 of the enzyme glycogenin. This is a reaction that is self-sustaining and requires the presence of magnesium ions in order to occur.

Glycogenin, with a molecular weight of 37,000, utilizes UDP-glucose as a substrate to autonomously elongate the glucan chain by six to seven glucosyl units through α (1→4) glycosidic bonds (Gardner et al., 2014). Glycogen synthase utilizes UDP-glucose to elongate the glucan primer of glycogenin. Glycogen synthase and primer glycogenin form a strong bond in a 1:1 ratio.

2.5. CARBOHYDRATE METABOLISM REGULATION

2.5.1. Intracellular Metabolic Regulators

The control point stages in the metabolic pathways for carbohydrates effectively regulate themselves by responding to molecules that reflect the condition of the cell.

The uptake of glucose is controlled by the availability of glucose-6-phosphate, as it inhibits the activity of hexokinase. The sole regulatory mechanism for glucose entry into "normal cells" is the transformation of glucose into glucose-6-phosphate. Typically, the rate of glucose transport exceeds the rate of glucose conversion. (It should be noted that skeletal muscle and adipose tissues are unique in that the uptake of glucose is typically limited by the rate of transport, which is determined by the number of transporters in the plasma membrane. The liver is a unique case as the rate of glucose intake increases with concentration, and the hepatic glucose transporter has a low affinity for glucose.

The glycolytic process is regulated by the energy state of the cell. According to Dashty (2013), cells with enough energy experience inhibition of phosphofructokinase and pyruvate kinase due to elevated ATP levels. In addition, increased NADH concentrations hinder the activity of isocitrate dehydrogenase, leading to an increase in citrate levels and the inhibition of phosphofructokinase. Both isocitrate dehydrogenase and citrate synthase are hindered by NADH. However, due to its higher susceptibility to NADH, isocitrate dehydrogenase is inhibited before citrate synthase. Finally, Phosphoenolpyruvate also hinders the activity of phosphofructokinase (Dashty, 2013). Phosphofructokinase regulators also control the levels of fructose-2,6-bisphosphate (AMP at high concentrations increases fructose-2,6-bisphosphate levels, while ATP at high levels decreases it).

The enzyme responsible for the production and degradation of fructose-2,6-bisphosphate controls the concentration of this regulatory substance. Typically, only one of the protein's two active sites is active at any given moment. The activity of phosphofructokinase decreases when the concentration of fructose-2,6-bisphosphate decreases, which is a result of elevated levels of ATP (Dashty, 2013).

As cellular energy levels decrease, AMP levels rise. AMP stimulates pyruvate kinase and phosphofructokinase. Furthermore, it stimulates the production of fructose 2,6-bisphosphate, leading to an increase in phosphofructokinase activity. Pyruvate kinase is additionally enhanced by fructose-1,6-bisphosphate; in the presence of an active phosphofructokinase, pyruvate kinase is activated by the product of the former. In most normal cells, glucose synthesis is negligible. The majority of non-liver cells contain moderate levels of phosphoenolpyruvate carboxykinase and pyruvate carboxylase. The impact of these substances on the regulatory enzyme, glucose-6-phosphate dehydrogenase, determines the relative levels of NADPH and NADP, which in turn govern the amount of glucose that passes through the hexose monophosphate pathway (Dashty, 2013).

Essentially, glycogen has a minimal impact on the function of most "typical cells." When the levels of cellular energy are high, indicated by elevated levels of ATP and glucose-6-phosphate, glycogen synthesis occurs in the cells that are capable of producing it. Conversely, glycogen breakdown occurs when the ratio of AMP to ATP increases.

2.5.2. Exogenous Metabolic Regulators

Like other cells, the liver and skeletal muscle autonomously regulate the metabolism of carbohydrates. However, these cell types must also possess the ability to alter their carbohydrate metabolism in response to external signals. In most metabolic scenarios, these two tissues, along with adipose tissue, which primarily participates in lipid metabolism, serve as the primary controllers of nutrient levels in circulation (Dashty, 2013).

The regulation has two primary goals: firstly, to ensure the maintenance of normal blood glucose levels despite changing circumstances, and secondly, to encourage physical activity as necessary.

The levels of adrenal hormones cortisol and adrenaline, pancreatic hormones insulin and glucagon, and neural impulses that drive muscular contraction exert significant regulatory influence over these tissues. The events described here streamline several complex procedures. Except for the mitochondrial enzyme pyruvate dehydrogenase, which is stimulated by calcium to remove phosphate, glucagon, adrenaline, and insulin typically induce the phosphorylation of regulatory enzymes. However, insulin causes the removal of phosphate. Several hormones, such as cortisol, insulin, and glucagon to a lesser degree, influence the quantity of enzymes within the cell. Enzyme concentration changes occur gradually, whereas phosphorylation and dephosphorylation occur rapidly (Dashty, 2013).

2.6. ANIMAL DISORDERS RELATED TO THE METABOLISM OF CARBOHYDRATES

The metabolism of glucose, fructose, and galactose is closely interconnected due to interactions across multiple biochemical pathways. Disorders that impact these processes can result in symptoms that vary in intensity from moderate to severe, and in some cases, can even be life-threatening. The clinical presentation encompasses various combinations of hepatomegaly, myalgia, and hypoglycemia. Sundrum (2015) suggests that specific dietary interventions can be employed to effectively treat or manage the majority of these illnesses.

2.6.1. Galactose and Fructose Disorders

Galactosemia is typically caused by a malfunctioning component of the second significant stage in the metabolic process of the sugar galactose. Galactose-1-phosphate accumulates as a result of galactose consumption, such as through the intake of milk. Consequently, the clinical symptoms of galactosemia become evident when milk feeding is started. If the feeding is not stopped, infants with the disorder will undergo weight loss, gradual liver failure, jaundice, lethargy, and kidney disease. In addition, they are susceptible to severe bacterial infections, specifically those caused by Escherichia coli (Sundrum, 2015).

2.6.2. Glycogen Storage Disorders

The brain, red blood cells, and the adrenal medulla rely on a continuous supply of glucose for their metabolic functions. The small intestine is the origin of this supply, where transport proteins aid in the assimilation of glucose into the cells lining the intestinal wall (Sundrum, 2015). Subsequently, glucose is transported into the bloodstream and ultimately reaches the liver, where it is stored in the form of glycogen. During periods of fasting, hunger, or other similar circumstances, the body converts glycogen into glucose and releases it into the bloodstream to provide a rapid source of energy. Furthermore, during physical exertion, the glycogen reserves in muscle tissue may undergo degradation. Glycogen storage disorders (GSD) can occur when the enzymes responsible for glycogen breakdown are hindered, resulting in the accumulation of glycogen in the muscle or liver. These disorders can affect the muscles, liver, or both, depending on the specific enzyme that is affected. Hypoglycemia occurs due to a malfunction in the last phase of glucose release by the liver in GSD type I, also referred to as von Gierke disease (Sundrum, 2015).

2.6.3. Congenital Disorders of Glycosylation (CDG)

Congenital disorders of glycosylation (CDG), also referred to as carbohydrate-deficient glycoprotein syndrome, have a significant impact on the brain and various other organs. The N-glycosylation pathway, occurring in the cytoplasm and endoplasmic reticulum, two cellular organelles responsible for protein and lipid synthesis, is where the primary biochemical abnormalities in CDG are located (Sundrum, 2015). Type I congenital disorders of glycosylation (CDG) are primarily caused by defects in phosphomannomutase 2, an enzyme involved in the processing of mannose.

During the early phases of lactation, cows experience negative energy balance (NEB) as a result of a sudden increase in their requirement for nutrients to enhance milk production, which exceeds the rise in their food consumption. Consequently, there is a decrease in blood glucose levels and the body's physiological reserves are mobilized to generate additional energy. In addition, there are decreased levels of calcium and phosphorus, as well as elevated blood concentrations of ß-hydroxybutyrate (BHBA) and non-esterified fatty acids (NEFA). According to Drackley, the phenomenon of uncontrolled mobilization of lipids in dairy cattle due to high negative energy balance (NEB) during early lactation is the cause of several health problems (Sundrum, 2015). All of these metabolic alterations increase the risk for ketosis, hepatic lipidosis, hypocalcemia, and infectious illnesses such as metritis and mastitis. However, there is a notable fluctuation in the plasma concentration of substrates during the day (diurnal variation) and over several days in the initial phases of breastfeeding.

There are notable disparities in the plasma levels between cows that did not experience clinical ketosis in the first six weeks of lactation and cows that exhibited metabolic issues, as indicated by their plasma BHBA concentration, which is commonly used as a marker for ketosis. The postpartum phase in animals housed in identical and carefully controlled environments on a research farm demonstrates significant fluctuations in the levels of hormones and metabolites. These variations indicate that individual cows' abilities to manage metabolic stress vary considerably (Sundrum, 2015).

2.7. THE SIGNIFICANCE OF CARBOHYDRATES IN MAINTAINING ANIMAL HEALTH

Optimal nutrition is crucial for maintaining the health and productivity of livestock in farming operations. Proper nutrition can enhance both the rate at which animals gain weight and their feed efficiency. Animal diets must be customized to meet their specific nutritional needs. If the animals' needs are not adequately met, they will be unable to grow, reproduce, or lactate, and they may even die.

Carbohydrates are composed of carbon, hydrogen, and oxygen and serve as feed components that supply energy. According to Knudsen et al. (2012), approximately 75% of an animal's diet should consist of them. Their energy enables the muscles to contract. Moreover, the metabolic energy generated by carbohydrates helps maintain the animal's body temperature. They enhance the body's utilization of lipids and amino acids. Carbohydrates are not stored in the body. The animal's diet must incorporate them daily. Unused carbohydrates undergo a process of transformation into fat and are subsequently stored.

Carbohydrates can be classified into two categories: simple and complex. Simple carbohydrates are rapidly digested. Simple carbohydrates consist of sugars and starches. Concentrates such as corn, wheat, oats, barley, and sorghum are composed of simple carbohydrates. Complex carbohydrates are more difficult to digest than simple carbohydrates. Lignin and cellulose are examples of complex carbohydrates. Most of these chemicals are present in pasture plants and hay. In addition, carbohydrates can be categorized into three distinct groups: sugars, starches, and fibers.

Sugars can be classified into two types: monosaccharides, which are also referred to as simple sugars, and disaccharides, commonly known as double sugars. Fructose, galactose, and glucose are examples of simple sugars. The three types of double sugars are lactose, sucrose, and maltose. Glucose is an excellent energy source for most cells (Knudsen et al., 2012).

1. **Starch:** It is a notable source of energy. Starch undergoes a process of conversion into glucose during digestion.
2. **Fiber:** It refers to the residual material left behind after the process of food digestion. Fiber is composed of cellulose and plant cells. Cellulose is the primary component of plant cell walls. Fiber provides volume, absorbs moisture, and is crucial for the proper functioning of the digestive system. Fiber promotes bacterial proliferation in the rumen, thereby playing a crucial role in ruminant digestion.
3. **Lipids:** These are synonymous with fats and oils. A lipid is a macronutrient that serves as a source of energy in food and is used by animals to store energy. Ether, a transparent solvent employed in nutritional studies, can be utilized for the purpose of dissolving lipids. Oils are in a liquid state and fats are in a solid state at typical temperatures. Fats contain the greatest amount of energy per unit of weight. Lipids actually possess 2.25 times the energy content of carbohydrates. An animal requires fats for the energy necessary for regular bodily maintenance. Fish meal and meat and bone meal are excellent sources of fats in animal diets.

Carbohydrates typically constitute 60–70% of the total energy content in the diets of most animals, serving as the primary source of energy. Carbohydrates not only provide the animal with its primary source of energy, but they also act as a foundation for the growth of ruminal bacteria and the production of microbial protein (Knudsen et al., 2012). Fibrous carbohydrates also promote optimal gastrointestinal function. The composition of carbohydrates in animal feeds, which includes both complex polymers and simple sugars, is a diverse blend that is commonly categorized according to the animal's ability to digest each type of carbohydrate. The insoluble neutral detergent fiber (NDF) and soluble neutral detergent fractions are two primary classifications of carbohydrates based on their solubility in neutral detergent. There are two additional classifications for NDF carbohydrates: forage NDF (fNDF) and nonforage NDF (nfNDF). Both of these classifications include cellulose, hemicellulose, and lignin. Neutral detergent-soluble carbohydrates (NDSC) encompass starch, water-soluble carbohydrates (WSC) like fructans and sugars, and neutral detergent-soluble fiber (NDSF) such as pectic compounds.

The availability of carbohydrates for an animal is determined by factors such as the origin and composition of the carbohydrates (e.g., growth conditions, type of food, processing methods). The ability of cattle to break down starch, lactose, microbial glycogen, and trehalose (a type of sugar) relies on the enzymes present in their small intestine lining and pancreatic secretions. In order for gastrointestinal microorganisms to provide nutrition to the animal, they need to break down and utilize all other carbohydrates, including sucrose and those present in NDF.

The digestibility of the carbohydrate components affects the animal's overall nutritional intake, while the fermentation in the rumen affects the type and timing of energy supply. This, in turn, affects the production of microbial proteins and the animal's energy intake and distribution. The type and origin of carbohydrates, as well as their interactions with other dietary components and animal factors, affect the type and timing of energy delivery (Knudsen et al., 2012). Physically effective NDF (peNDF) refers to the portion of NDF in the diet that contains particles of sufficient size to form a rumen mat. This mat serves to capture small, potentially biodegradable particles in the rumen and encourages rumination. The longer period of time that digesta remains in the rumen not only improves total-tract digestibility and increases buffering capacity, but also reduces the risk of low ruminal pH and a decrease in NDF digestibility (Knudsen et al., 2012).

Chapter **3**

Lipid Metabolism in Animals

Contents

Lipid metabolism refers to the intricate cellular process of synthesizing and degrading lipids, also known as fats. It involves synthesizing essential lipids for the construction of cell membranes and also storing fats for energy. Animals synthesize these lipids in the liver as well as acquire them from their diet. Lipid digestion, absorption, and lipogenesis are significant factors. This chapter will provide a thorough examination of lipid metabolism in animals, focusing on the structure and function of lipids, the metabolism of fatty acids, cholesterol metabolism, lipoprotein metabolism, the regulation of lipid metabolism, disorders associated with lipid metabolism, and the role of lipids in animal health.

3.1. LIPID STRUCTURE AND FUNCTION

Lipids and fats are essential for maintaining homeostasis in the human body. Lipids (Figure 3.1) play a crucial role in facilitating several vital functions of the body. Lipids refer to substances that are characterized by their fatty, waxy, or oily nature. According to Egom & Hafeez (2016), these substances can dissolve in organic solvents but cannot dissolve in polar solvents such as water.

Among lipids are:

- Triglycerides, which are commonly known as fats and oils.
- Phospholipids;
- Wax and steroids.

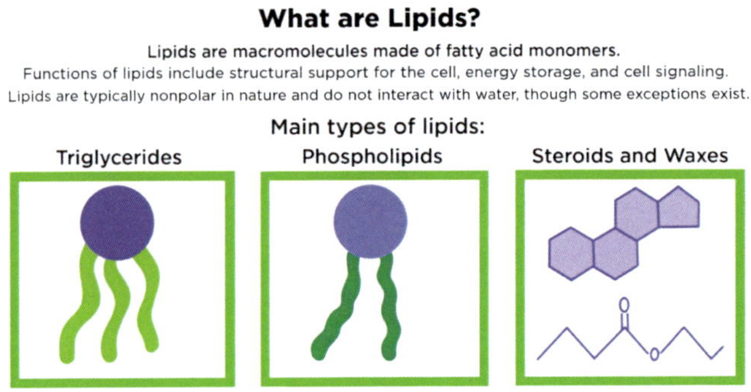

Figure 3.1.*Lipids are diverse macromolecules with various functions in cells.*

Source: Slizewska (2020).

Fats and oils are composed of esters of glycerol, which is a polyol consisting of three carbons, and three fatty acids. Fatty acids are hydrocarbon chains that have different lengths and levels of saturation, and they end with carboxylic acid groups. In addition, there exist multiple types of fatty acids that can be generated through cis or trans fatty acid double bonds. Fatty acids in biological systems generally consist of 14 to 24 carbon atoms and possess an even number of carbon atoms. Triglycerides facilitate the assimilation of fat-soluble vitamins, serve as an energy reservoir, and provide cellular insulation. Fats generally exist in a solid state at room temperature, whereas oils are typically in a liquid state (Egom & Hafeez, 2016).

Lipids are a crucial component of the cell membrane. The structure typically comprises a hydrophobic fatty acid tail, a phosphate group, and a hydrophilic glycerol backbone. Phospholipids are therefore amphipathic. Phospholipids are arranged in a double layer in the cell membrane, serving as a protective barrier against certain chemicals. The hydrophobic segment is oriented towards the interior, while the hydrophilic segment is oriented towards the exterior. This configuration facilitates monitoring the ingress and egress of specific molecules in and out of the cell. For example, nonpolar molecules and small polar molecules, such as oxygen and water, can easily diffuse into and out of cells. Transport proteins facilitate the efficient passage of large, polar molecules such as glucose (Egom & Hafeez, 2016).

Steroids belong to a distinct category characterized by the presence of four fused rings. Cholesterol is a prominent type of steroid. The liver synthesizes cholesterol, which serves as the precursor for various steroid hormones such as cortisol, estrogen, and testosterone. It penetrates the bilayer of cell membranes and alters the fluidity of the membrane by inserting itself into it (Feingold, 2024).

The interaction between behaviors that favor fat and behaviors that avoid water becomes more apparent during the transportation of lipids in the plasma. Triglycerides and cholesterol are lipid molecules that lack polarity. Therefore, they require the aid of lipoprotein particles to traverse the polar plasma. The main role of lipoproteins is to enable the transportation of hydrophobic lipids in aqueous environments. Lipoprotein structure consists of triglycerides, cholesterol, phospholipids, and apolipoproteins. Apolipoproteins primarily function as carrier proteins, facilitating the exchange of lipid components between lipoproteins. Additionally, they serve as cofactors for enzymes involved in the metabolism of lipoproteins (Feingold, 2024). Chylomicrons, LDL, HDL, IDL, and VLDL are different types of lipoproteins. Each molecule is utilized in a specific phase of lipid transportation (Feingold, 2024).

Chylomicrons are large particles that are rich in triglycerides and are produced by the endoplasmic reticulum of enterocytes in the small intestine. They play a role in the transportation of cholesterol and triglycerides from food to the liver and other peripheral tissues (Feingold, 2024). Apolipoprotein Apo B-48 has a vital role in the process of absorbing dietary lipids and fat-soluble vitamins. It accomplishes this by participating in the assembly of chylomicrons. Very low-density lipoproteins (VLDLs), which are high in triglycerides, are synthesized in the liver. Apo B-100 plays a vital role in the production of VLDL. Muscle and adipose tissue extract triglycerides from VLDL to create IDL particles that are high in cholesterol.

VLDL and IDL particles merge to produce LDL particles, which are rich in cholesterol as well (Feingold, 2024). LDL, commonly known as "bad cholesterol," carries the majority of cholesterol in the bloodstream. Apo B-100 is essential for the liver and other organs to absorb LDL particles through the LDL receptor-mediated process (Feingold, 2024). HDL particles, which are abundant in phospholipids and cholesterol, facilitate the transportation of cholesterol from peripheral tissues to the liver for elimination. HDL cholesterol is commonly referred to as "good cholesterol."

There is only one external pathway for transporting dietary triglycerides and cholesterol from the small intestine. Triglycerides undergo hydrolysis in the small intestine through the action of bile acids, such as cholic acid, and enzymes. The duodenum secretes the hormone cholecystokinin (CCK) in response to initial digestion products such as free fatty acids. Furthermore, CCK activity stimulates the pancreas to secrete pancreatic digestive enzymes into the gastrointestinal tract, while also promoting the contraction of the gallbladder, resulting in the release of bile into the small intestine (Feingold, 2024). Bile acids possess a detergent property that facilitates the emulsification of fats, thereby increasing their surface area and enabling water-soluble digestive enzymes to more easily break them down through hydrolysis. Through the utilization of mixed micelles generated during the process, pancreatic lipase, a notable enzyme, decomposes triglycerides into free fatty acids and monoacylglycerol. These substances are then assimilated by the cells lining the intestinal mucosa (Feingold, 2024).

The endogenous system is an alternative route through which triglycerides and cholesterol are transported from the liver and other non-intestinal tissues into the bloodstream. The liver synthesizes triglycerides from free fatty acids and carbohydrates. The VLDL core subsequently releases these triglycerides into the bloodstream. The VLDL particles and lipoprotein lipase interact in tissue capillaries, leading to the hydrolysis of the triglyceride core and the liberation of free fatty acids. After being extracted from plasma, some of the residual particles adhere to hepatic cells. However, the residual particles undergo a transformation into LDL particles, which serve as a source of cholesterol for cells such as the kidneys, skeletal muscle, adrenal glands, gonads, and lymphocytes that possess LDL receptors. In addition to the previously mentioned functions, fat can also be metabolized for energy when necessary. Adipose triglyceride lipase

(ATGL), hormone-sensitive lipase (HSL), and monoglyceride lipase (MGL) are stimulated to release fatty acids through the action of adrenaline during exercise or glucagon during fasting. Most tissues can utilize these fatty acids as a source of energy through the assistance of mitochondria and the Krebs cycle (Feingold, 2024).

3.2. ANIMAL FATTY ACID METABOLISM

Fatty acid metabolism refers to a set of metabolic processes that involve or are closely related to the family of molecules known as fatty acids, which are a type of macronutrient in lipids. These can be categorized into two main groups: (1) energy-producing catabolic processes; and (2) anabolic processes, which utilize them as fundamental components for the creation of other molecules (Wakil & Abu-Elheiga, 2009).

During catabolism, fatty acids are metabolized to produce energy, mainly in the form of adenosine triphosphate (ATP). Fatty acids, when completely metabolized through beta oxidation and the citric acid cycle, yield the greatest ATP production per gram compared to other macronutrients such as protein and carbohydrates. Thus, in most animals, fatty acids, primarily in the form of triglycerides, serve as the main type of energy storage.

Intact fatty acids have a crucial function in the anabolic process by serving as building blocks for phospholipids, hormones, triglycerides, second messengers, and ketone bodies. Phospholipids form the phospholipid bilayers that serve as the foundation for all of the cell's membranes. Fatty acids are utilized in the production of these membranes (Wakil & Abu-Elheiga, 2009). The phospholipids constitute the plasma membrane and other membranes that house various organelles within cells, such as the Golgi apparatus, mitochondria, endoplasmic reticulum, and nucleus. Fatty acids undergo a distinct form of anabolism to produce various substances such as second messengers and local hormones. Among these local hormones, prostaglandins are likely the most widely recognized. They are produced from arachidonic acid, which is stored in the cell membrane (Wakil & Abu-Elheiga, 2009).

3.2.1. Metabolism of Fatty Acids

Triglycerides, also known as fat stores, serve as the storage site for fatty acids in adipose tissue. They are released intermittently during the intervals between meals.

Lipases catalyze the hydrolysis of triglycerides, liberating fatty acid chains from the glycerol backbone in which they are stored. The activation of these lipases is induced by increased blood concentrations of adrenaline and glucagon (or norepinephrine released by sympathetic neurons in adipose tissue). This increase is a result of a decrease in blood glucose levels after meals, which consequently leads to a decrease in blood insulin levels (Wakil & Abu-Elheiga, 2009).

Once released from glycerol, the unbound fatty acids are transported in the bloodstream and distributed throughout the body by plasma albumin.

Transport proteins, specifically the SLC27 family fatty acid transport protein, facilitate the entry of long-chain free fatty acids into metabolizing cells. These cells are found in most parts of the body, except for red blood cells and neurons in the central nervous system. The tissues of the central nervous system are unable to utilize fatty acids, despite having mitochondria, due to the fact that long-chain fatty acids, unlike medium-chain fatty acids, are unable to pass through the blood-brain barrier and enter the interstitial fluids surrounding these cells. Red blood cells cannot metabolize fatty acids due to their lack of mitochondria.

Upon entering the cell, a fatty acid molecule combines with ATP (which is converted to AMP and inorganic pyrophosphate) to form a fatty acyl-adenylate. This compound then reacts with free coenzyme A to produce an acyl-CoA molecule.

The carnitine shuttle is necessary for the transportation of acyl-CoA into the mitochondrion. Carnitine palmitoyl transferase I is located on the cytosolic surfaces of both the outer and inner mitochondrial membranes. Its function is to transfer acyl-CoA to the hydroxyl group of carnitines.

A carnitine-acylcarnitine translocase facilitates the movement of acyl-carnitine into the cell, while a carnitine transporter moves carnitine out of the cell.

Carnitine palmitoyl transferase II converts acyl-carnitine back into acyl-CoA on the inner surface of the inner mitochondrial membrane. One shuttle transports the released carnitine back to the cytosol, while another shuttle carries the acyl-CoA into the mitochondrial matrix.

The fatty acids, which are in the form of acyl-CoA molecules, undergo beta oxidation in the mitochondrial matrix. This process breaks down the long carbon chains of the fatty acids into a series of two-carbon units called acetate. When co-enzyme A is combined with these units, it results in the formation of acetyl CoA molecules. These molecules then combine with oxaloacetate to produce citrate at the initiation stage of the citric acid cycle (Wakil & Abu-Elheiga, 2009). The addition of acetyl-CoA to the cycle results in the release of CO_2 and H_2O, along with a substantial amount of energy that is stored as ATP. This reaction can be considered as the initial step of the cycle.

Essentially, beta oxidation occurs in the following manner:

- Acyl-CoA dehydrogenase undergoes dehydrogenation, resulting in the production of one molecule of $FADH_2$.
- Hydration mediated by enoyl-CoA hydratase.
- The enzyme 3.3-hydroxyacyl-CoA dehydrogenase catalyzes the oxidation of a substrate, resulting in the production of 1 molecule of $NADH^+ H^+$.

Thiolase catalyzes the cleavage reaction, resulting in the formation of one molecule of acetyl-CoA and a fatty acid that has been shortened by two carbon atoms, leading to the production of a new, shorter acyl-CoA (Wakil & Abu-Elheiga, 2009).

The beta oxidation reaction continues until the fatty acid is completely converted into acetyl-CoA. In the case of fatty acids with an odd number of carbon atoms, one molecule of propionyl-CoA is also produced along with each molecule of fatty acid. During oxidative phosphorylation, each beta oxidation cleavage of the acyl-CoA molecule ultimately generates five ATP molecules.

The citric acid cycle of the mitochondrion is initiated by the acetyl-CoA produced through beta oxidation. This acetyl-CoA then combines with oxaloacetate to form citrate. When oxidative phosphorylation is combined with it, acetyl-CoA undergoes complete combustion, resulting in the production of CO_2 and water. Each oxidation of an acetyl-CoA molecule results in the release of 1 GTP and 11 ATP molecules, which are subsequently collected. Acetyl-CoA is the final product of fatty acid beta oxidation, except in certain liver conditions. Subsequently, propionyl-CoA undergoes conversion into succinyl-CoA through the action of the enzymes methylmalonyl-CoA mutase (MCM), which relies on vitamin B12, and propionyl-CoA carboxylase (PCC), which relies on biotin. Before entering the citric acid cycle, succinyl-CoA undergoes a series of transformations to become malate and then pyruvate. Subsequently, it is transferred to the matrix.

According to Wakil & Abu-Elheiga (2009), the liver is the exclusive site where this reaction can occur. In the liver, glycerol kinase phosphorylates the glycerol generated by lipase activity, leading to the formation of glycerol 3-phosphate, which is then oxidized to dihydroxyacetone phosphate. The glycolytic enzyme triose phosphate isomerase converts this substance into glyceraldehyde 3-phosphate. Glyceraldehyde 3-phosphate can undergo oxidation through glycolysis or be converted to glucose through gluconeogenesis (Wakil & Abu-Elheiga, 2009).

3.3. ANIMAL CHOLESTEROL METABOLISM

**** Fig/Structure of cholesterol include**

Cholesterol metabolism in animals occurs in multiple organs. Approximately 90% of the unbound cholesterol present in animal cells is located within the plasma membrane. Cholesterol is utilized by the body for various essential processes, such as the production of steroid hormones in specific cells (Gérard, 2013). In addition, bile acids, which are synthesized by the liver and subsequently transported to the stomach to facilitate the breakdown of dietary fat, serve as precursors to cholesterol. Unlike most other biological components, cholesterol is not metabolized by individual cells. Instead, the levels of cholesterol in cells are regulated by a complex interplay between the production of cholesterol within the body, the absorption of cholesterol from outside sources, and the release of cholesterol from within cells into the bloodstream. High levels of cholesterol lead to the development of plaques in the walls of important arteries, resulting in a condition known as atherosclerosis (Gérard, 2013).

Cholesterol, which is enclosed within lipids and proteins as cholesterol esters, is transported throughout the body. Cholesterol esters are significantly more hydrophobic than cholesterol due to the presence of a fatty acid esterified to the hydroxyl group. The liver, being the main site of cholesterol synthesis (Figure 3.2) in animals, ultimately takes up the chylomicrons formed from dietary cholesterol by the intestine. Subsequently, these particles are transported via the circulatory system. LDLs, also known as low-density lipoproteins, are cholesterol that is produced by the liver and packaged for transport to different organs through the bloodstream. LDL particles are internalized by other cells through receptor-mediated endocytosis and subsequently transported to lysosomes via the endocytic pathway (Gérard, 2013). Most of the cholesterol derived from free LDL is transported back to the plasma membrane by a cytoplasmic carrier protein once cholesterol esters are broken down within the lysosome. Importantly, a portion of cholesterol is transferred to the extracellular reservoir (ER). The level of fat in this reservoir plays a crucial role in activating transcription factors that control genes associated with cholesterol metabolism (Gérard, 2013).

The presence of sterol-sensing domains on two crucial enzymes in the endoplasmic reticulum allows them to respond to the amount of cholesterol in the cell membrane and control the levels of free cholesterol within the cell. The enzyme acyl coenzyme A cholesterol acyltransferase (ACAT) is responsible for converting free cholesterol into cholesterol esters, which are then stored. This enzyme is activated when there is an accumulation of cholesterol derived from LDL in the endoplasmic reticulum (ER). Elevations in free cholesterol or its oxygenated metabolite lead to the degradation of the enzyme β-hydroxy-βmethylglutaryl-coenzyme A (HMG-CoA) reductase. This enzyme is responsible for catalyzing the first step in cholesterol production. The degradation of HMG-CoA reductase is facilitated by cholesterol through a pathway that relies on proteasomes and ubiquitin.

In addition, there is a third protein known as SREBP cleavage activating protein (SCAP) that contains a sterol-sensing domain. This protein captures the sterol regulatory element-binding protein (SREBP), which is responsible for activating transcription, and keeps it trapped in the endoplasmic reticulum (ER). This limits the manifestation of the LDL receptor and HMG-CoA reductase genes. When the levels of cholesterol in the endoplasmic reticulum (ER) are low, the SCAP/SREBP complex moves to the Golgi apparatus. At the Golgi apparatus, the complex undergoes proteolytic cleavage, which leads to the release of the SREBP activation domain. Subsequently, it translocates to the nucleus, where it induces the synthesis of proteins implicated in cholesterol metabolism, such as HMG-CoA reductase and the LDL receptor (Gérard, 2013). This negative feedback mechanism reduces the intake of cholesterol from external sources as well as the production of cholesterol through de novo synthesis.

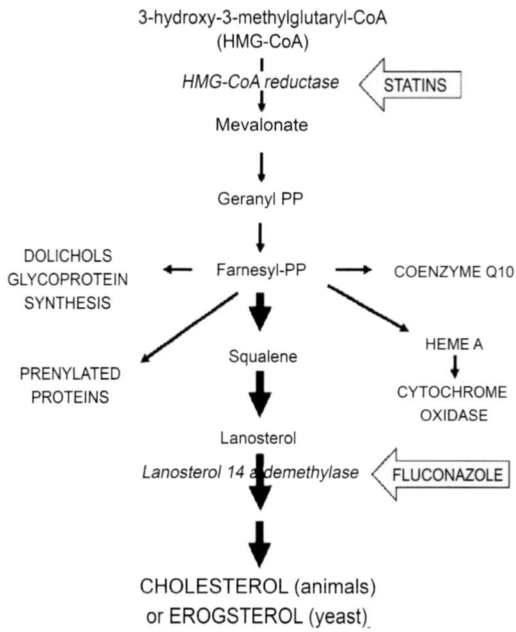

Figure 3.2. *Simplified schematic view of cholesterol and ergosterol biosynthesis pathway in animals and yeast.*

Source: Dhakal et al. (2019).

3.4. ANIMAL LIPOPROTEIN METABOLISM

Lipoproteins consist of lipids and proteins, specifically known as apolipoproteins (Phillips, 2014). The diagram below illustrates the four primary pathways involved in lipoprotein metabolism. The process of delivering dietary and hepatic-derived lipid to peripheral cells can be conceptualized as lipid absorption through exogenous and endogenous pathways. All of them rely on lipoprotein particles that contain apo B. The three pathways mentioned are crucial for the energy metabolism process as they transport fatty acids, which are generated during the breakdown of triglycerides and, to a lesser extent, cholesteryl esters on lipoproteins, to cells in the periphery (Phillips, 2014). Concerning the pathophysiology of atherosclerosis, these three mechanisms ultimately result in the overall delivery or transportation of cholesterol to cells in the periphery. If vascular wall cells also accumulate, it can lead to the development of atherosclerosis. Peripheral cells are prone to cholesterol accumulation due to their inability to break down cholesterol through enzymatic pathways, which are present in liver cells. Additionally, peripheral cells also produce cholesterol themselves. Furthermore, due to its water insolubility, cholesterol cannot readily diffuse away from the location where it is produced or deposited (Phillips, 2014).

Peripheral cells primarily depend on the HDL-mediated reverse cholesterol transport system to regulate their cholesterol levels. This process facilitates the transportation of excess cholesterol from cells in the periphery back to the liver, where it can be eliminated either as free cholesterol or after conversion into bile acids. Hence, the liver plays a role in both the facilitation and reversal of cholesterol transport pathways, serving as a buffer to help the animal's body maintain cholesterol homeostasis (Phillips, 2014).

3.4.1. Process of Lipid Absorption

Due to their hydrophobic nature, lipids require specific mechanisms to facilitate the absorption of the 60–130 g of fat typically consumed daily in a Western diet, allowing them to pass through the intestinal wall more efficiently. During the process of digestion, pancreatic lipase converts dietary lipids into molecules that are more polar and amphipathic by removing fatty acids (Phillips, 2014). Consequently, cholesterol esters are converted into free cholesterol, phospholipids are transformed into lysophospholipids, and triglycerides are broken down into monoglycerides and diglycerides. Within the intestinal lumen, these lipids with both hydrophilic and hydrophobic properties combine with bile acids to form large aggregates called micelles. Lipid absorption occurs when the micelles come into contact with the microvillus membranes of the intestinal mucosal cells. Some lipids can be absorbed through passive transfer, but recent studies suggest that specific transporters, like the NPC1L-1 transporter for cholesterol, may also play a role in absorption. Short-chain free fatty acids are readily taken up by the portal circulation and transported to the liver via albumin if they contain 10 or fewer carbon atoms. The intestinal cells convert the long-chain fatty acids, monoglycerides, and diglycerides that are consumed into triglycerides and cholesteryl esters through a process called reesterification. The microsomal transfer protein subsequently encapsulates the recently synthesized triglycerides and cholesteryl esters, together with apo B-48, within chylomicrons (Phillips, 2014).

3.4.2. Exogenous Pathway

The newly synthesized chylomicrons in the gastrointestinal tract are initially released into the lacteals, then they pass through the lymphatic ducts and finally enter the bloodstream via the thoracic duct. Chylomicrons engage with proteoglycans on the inner surface of capillaries in different tissues, such as skeletal muscle, the heart, and adipose tissue, upon entering the bloodstream. These proteoglycans contain heparin sulfate. Proteoglycans and GPIHBPI in capillaries aid in the attachment of LPL,20, which breaks down triglycerides on chylomicrons (Phillips, 2014).

Cells can utilize the liberated fatty acids and glycerol generated by the hydrolysis of triglycerides by LPL as a source of energy. In order to store energy for an extended period of time, surplus fatty acids are converted back into triglycerides, primarily within fat cells known as adipocytes. Apo C-II, a protein found on VLDL, plays a crucial role in the activation of LPL and is a key factor in triglyceride metabolism. Hormone-sensitive lipase, an additional enzyme, is situated within adipose cells. During periods of fasting, when the body demands more energy than carbohydrates can supply, this enzyme liberates free fatty acids from triglycerides stored in fat deposits (Phillips, 2014). The hormone cortisol and adrenaline facilitate the release and hydrolysis of triglycerides from adipocytes. Conversely, insulin, when taken immediately after a meal, inhibits adipocyte lipolysis and promotes the storage of fat and the utilization of glucose (Phillips, 2014).

3.4.3. Endogenous Pathway

The liver obtains the majority of triglycerides packed into VLDL from the diet after they circulate from adipose tissue. Typically, the liver only produces a small amount of carbohydrates from the ones consumed. Once VLDL particles are introduced into the bloodstream, they undergo a lipolytic process similar to that of chylomicrons. When the core lipids of VLDL are lost, it transforms into denser particles called IDLs or VLDL remains. This causes the apolipoproteins and phospholipids to separate and transfer to other lipoprotein particles, such as HDL. This transfer is primarily facilitated by the action of LPL (Phillips, 2014).

The cholesteryl ester transport protein facilitates the exchange of triglycerides for cholesterol esters from HDL, enabling the temporary existence of IDL. Like chylomicron remnants, IDL is taken up by the liver via apo E. Hepatic triglyceride lipase, located on hepatic endothelial cells, subsequently metabolizes the triglyceride in intermediate-density lipoprotein (IDL) and the LDL receptor, resulting in the production of low-density lipoprotein (LDL). Over time, approximately 50% of VLDL is completely transformed into LDL, while the remaining portion is taken up by the liver's residual receptors as VLDL remnants (Phillips, 2014).

LDL particles, due to their high affinity for LDL receptors, are responsible for transporting exogenous cholesterol to peripheral cells. LDL particles can bind to LDL receptors located on cell membranes, which specifically recognize apo B-100. This binding allows LDL particles to traverse across capillary endothelial cells. Once the LDL receptors have bound to these molecules, they are internalized by cells through endocytosis and subsequently degraded within the lysosome. Acid lipase converts the triglycerides in LDL into glycerol and free fatty acids, which are subsequently utilized by the cell for energy metabolism. Alternatively, the triglycerides undergo reesterification and are stored in lipid droplets for future utilization. In the process of membrane biosynthesis, the free cholesterol obtained from the breakdown of LDL can be utilized. Any excess cholesterol is converted into cholesteryl esters and stored by an enzyme called acyl-CoA–cholesterol acyltransferase (ACAT). The presence of LDL receptors affects the control of cellular cholesterol synthesis (Phillips, 2014). Excessive cellular cholesterol triggers a sophisticated process that involves both gene regulation and posttranscriptional gene regulation. This process reduces the activity of the LDL receptor and several other enzymes in the cholesterol biosynthetic pathway, such as HMG-CoA reductase, which is the primary target of cholesterol-lowering statin-type drugs (Phillips, 2014).

3.4.4. Pathway of Reverse Cholesterol Transport

As previously stated, one of the main roles of HDL is to maintain the equilibrium of cholesterol levels in peripheral cells by utilizing the reverse cholesterol transport pathway. HDL is believed to remove excess cholesterol from cells through various mechanisms (Phillips, 2014). HDL effectively absorbs the small amount of cholesterol that is able to move out of the cells through the aqueous diffusion pathway. Cholesterol, being an amphipathic lipid, is able to dissolve in plasma at micromolar concentrations even though it is not very soluble in water. Additionally, it has the ability to autonomously detach from the surface of cellular membranes and migrate into the extracellular fluid. Once converted into cholesteryl ester in the extracellular space, certain free cholesterol molecules will then bind to developing HDL and become immobilized within lipoproteins (Phillips, 2014).

Initially, the developing high-density lipoprotein (HDL) undergoes a transformation process to become HDL3. The synthesis of cholesteryl ester through LCAT increases the ability of the HDL3 particle's surface to absorb more free cholesterol from cell membranes, as well as phospholipids from VLDL and chylomicrons and apo C-I, C-II, C-III, and E. This process leads to the conversion of HDL3 to HDL2 (Phillips, 2014). HDL2 can transport cholesterol to the liver through the scavenger receptor type B1 (SR-B1) and possibly other receptors. In addition, HDL particles can transfer cholesteryl esters to chylomicrons and VLDL remnants, which are then transported to the liver. The cholesteryl ester transfer protein (CETP) facilitates the transfer of cholesterol between HDL and LDL, connecting the pathways of forward and reverse cholesterol transport. Initially, CETP transfers cholesterol from HDL to LDL, and then the LDL receptor returns approximately half of the cholesterol from HDL back to the liver. Upon entering the liver, cholesterol is either directly expelled into the bile or undergoes conversion into bile acid before being expelled (Phillips, 2014).

3.5. REGULATION OF LIPID METABOLISM

Lipid metabolism refers to the cellular processes involved in the synthesis and degradation of lipids. The process involves the breakdown and storage of fats to produce energy, as well as the creation of structural and functional lipids, such as those required for constructing cell membranes (Birsoy et al., 2013). These fats are endogenously synthesized by the liver in animals and obtained from dietary sources. Lipogenesis is the process responsible for the production of these fats. The human body synthesizes the majority of its lipids from food consumption in the form of triglycerides and cholesterol. Fatty acids and membrane lipids are additional types of lipids found in the body. While the process of breaking down and absorbing dietary fat is commonly known as "lipid metabolism," organisms have two distinct sources of fat for obtaining energy: stored fat and fat obtained through diet. Lipids are nonpolar molecules, which means that they repel water. Therefore, in order for their metabolism to begin, they need to be dissolved. Hydrolysis, which is the initial stage of lipid metabolism, is facilitated by various enzymes found in the digestive system (Birsoy et al., 2013). The regulation of animal lipid metabolism involves intricate systems that ensure a harmonious equilibrium between lipid synthesis, breakdown, and storage. The processes involved in the regulation of lipid metabolism in animals include:

- Regulation of hormones;
- Enzymatic processes and metabolic pathways;
- Metabolism of lipids and their uptake into the body.

3.5.1. Hormonal Control of Lipid Metabolism

Hormones play a crucial role in regulating lipid metabolism (Figure 3.3) to ensure a proper equilibrium between energy storage and utilization.

- Insulin promotes lipid synthesis, inhibits lipid breakdown, and facilitates the transport of glucose while aiding its conversion into fatty acids.
- Glucagon reduces hepatic steatosis and VLDL output by inhibiting the production of new fatty acids and promoting the oxidation of fatty acids.

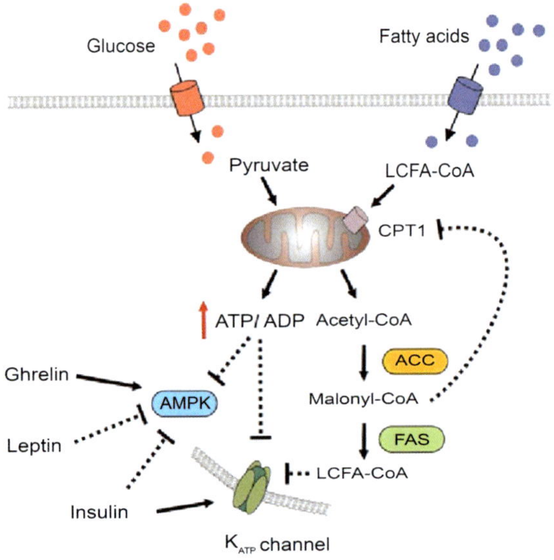

Figure 3.3. *Regulation of lipid metabolism by wild-type and mutant p53.*

Source: Jordan et al. (2010).

- The thyroid hormone promotes mitochondrial fatty acid β-oxidation, enhances lipolysis, boosts hepatic triglyceride production, and regulates cholesterol levels through various mechanisms, such as upregulating LDL receptors and promoting bile acid excretion.

- Estrogen may reduce the production of triglycerides and increase HDL cholesterol levels by promoting fatty acid oxidation and inhibiting lipogenesis.
- Testosterone initially stimulates the process of de novo lipogenesis and the creation of fat in order to store energy. These fats are then combined with triglycerides.
- Adrenaline impacts the process of lipid turnover and accumulation in adipose tissue by promoting the breakdown of lipids (lipolysis) and hindering the formation of new lipids (lipogenesis) through the activation of AMPK (adenosine monophosphate-activated protein kinase) (Birsoy et al., 2013).

3.5.2. Enzymes and Pathways

Beta oxidation is the metabolic process that breaks down lipids. It takes place in the peroxisome and mitochondrial cell organelles.

Triglycerides are liberated from the chylomicrons (or other lipoproteins) as they traverse the tissues and undergo degradation by lipoprotein lipase on the inner surface of endothelial cells in capillaries. Prior to entering cells, triglycerides undergo conversion into fatty acids and glycerol. Any remaining cholesterol is subsequently transported back to the liver through the bloodstream (Birsoy et al., 2013).

In the cytoplasm of a cell, such as a muscle cell, glycerol is converted into glyceraldehyde 3-phosphate, which is an intermediate molecule in the process of glycolysis. In this process, glyceraldehyde 3-phosphate is further oxidized to generate energy. Nevertheless, the mitochondria are primarily responsible for the majority of fatty acid breakdown. In order for long chain fatty acids (those containing more than 14 carbons) to permeate the mitochondrial membrane, they must undergo conversion into fatty acyl-CoA. Fatty acid catabolism initiates in the cytoplasm of cells when acyl-CoA synthetase catalyzes the addition of coenzyme A to the fatty acid, utilizing the energy released from ATP hydrolysis (Birsoy et al., 2013). Consequently, the acyl-CoA molecule traverses the mitochondrial membrane and initiates the beta oxidation process. The main products of the beta oxidation pathway are NADH, FADH, and acetyl-CoA. Acetyl-CoA is then used to produce energy in the citric acid cycle. The enzymes acyl-CoA dehydrogenase, enoyl-CoA hydratase, 3-hydroxyacyl-CoA dehydrogenase, and 3-ketoacyl-CoA thiolase are essential for the beta oxidation process. Acetyl-CoA is synthesized from fatty acids, as depicted in the left diagram. The complete net reaction, using palmitoyl-CoA (16:0) as a representative substrate, involves the consumption of 7 FAD molecules, 7

NAD^+ molecules, 7 CoASH molecules, 7 water molecules, and the production of $H(CH_2CH_2)$. The reaction $7CH_2CO\text{-}SCoA$ yields 8 $CH_3CO\text{-}SCoA$, 7 $FADH_2$, 7 NADH, and 7 H^+.

3.5.3. Lipid Digestion

The initial phase of lipid metabolism is referred to as digestion, which entails the utilization of lipase enzymes to decompose triglycerides into smaller monoglyceride units. Lingual lipase is the initial enzyme responsible for the chemical digestion of fats in the oral cavity. Lipases do not break down consumed cholesterol; instead, it remains unchanged until it reaches the epithelial cells of the small intestine.

Subsequently, lipids enter the stomach where gastric lipase further breaks down the food chemically, while peristalsis initiates mechanical digestion. Upon entering the small intestine, the majority of lipid digestion and absorption occurs there. Pancreatic lipase family and bile salt-dependent lipase are enzymes released into the small intestines to facilitate the mechanical digestion of triglycerides. They break down triglycerides into individual fatty acid units, which can then be absorbed by the epithelial cells of the small intestine. The pancreatic lipase is responsible for generating the signal that triggers the breakdown of triglycerides into separate free fatty acid and glycerol units (Birsoy et al., 2013).

3.5.4. Assimilation of Lipids

Fat absorption is the second stage in lipid metabolism. Although the majority of fat absorption primarily occurs in the small intestines, it is worth noting that short chain fatty acids can also be absorbed in the stomach (Birsoy et al., 2013). Upon hydrolysis, the triglycerides will aggregate into micelles, which are structures composed of individual fatty acids, glycerols, and cholesterol. Upon leaving the micelles, fatty acids and monoglycerides disperse over the membrane and penetrate the intestinal epithelial cells.

The process of recombining fatty acids and monoglycerides into triglycerides occurs within the cytoplasm of epithelial cells. Triglycerides and cholesterol are enclosed within larger particles called chylomicrons. These chylomicrons are structures that contain both hydrophilic and hydrophobic components, and they transport digested lipids. Chylomicrons are located in the cytoplasm of epithelial cells. Chylomicrons will be transported to the body's adipose and other tissues via the bloodstream (Birsoy et al., 2013).

3.6. DISORDERS OF LIPID METABOLISM IN ANIMALS

Abnormalities in the way animals process lipids can greatly affect their overall health and well-being. Lipids, being hydrophobic macromolecules, have significant functions in the body, such as storing energy and forming lipoprotein particles and cellular membranes. The cells lining the small intestine absorb lipids and convert them into lipoprotein particles. These particles then travel through the lymphatic system and ultimately reach the liver. Lipoproteins, including chylomicrons, LDL, HDL, IDL, and VLDL, transport fat-soluble vitamins, cholesterol, and triglycerides in the bloodstream.

Nevertheless, malfunctions in the enzymes responsible for metabolizing fats, the cell receptors that recognize various types of lipoproteins, or the structural proteins of lipoprotein particles can result in disorders that impact lipid metabolism. These disorders can cause lipids to build up in the walls of blood vessels, resulting in conditions such as atherosclerosis, which is characterized by abnormal thickening and hardening of artery walls.

Lipid metabolism disorders are characterized by dysregulation of lipid synthesis, catabolism, digestion, uptake, and transportation in the body. This can lead to an imbalance of lipids in various tissues, which can affect multiple physiological processes. The primary lipids present in blood are cholesterol, triacylglycerol (TAG), phospholipids (PL), and free fatty acids. The liver has a vital function in regulating lipid metabolism through the uptake of free fatty acids, which promotes the production, storage, and utilization of fatty acids.

The lipid-based plasma membrane serves as a protective barrier that separates the interior and exterior of cells, shielding them from oxidative harm induced by free radicals. Nevertheless, in the presence of oxidative damage, phospholipids and cholesterol esters that consist of polyunsaturated fatty acids (PUFAs) have the ability to undergo oxidation and generate oxidative metabolites. These compounds have the potential to harm membrane proteins, enzymes, or receptors, resulting in elevated membrane leakage and reduced membrane fluidity. Extended exposure to an environment abundant in oxidative damage can disturb lipid metabolism and impair its optimal functioning (Heindel et al., 2017).

Animals afflicted with lipid metabolism disorders can experience a range of diseases that hinder the processing of lipids. An instance of such disorders is lipid storage diseases, which belong to a category of inherited metabolic disorders resulting in the buildup of fats or lipids in the tissues and cells of the body. These diseases typically arise from the partial or complete lack of enzymes involved in metabolic processes. Animals afflicted with genetic storage diseases may exhibit an outwardly normal appearance upon birth, but will subsequently manifest clinical symptoms within the initial weeks or months of their existence. Regrettably, the majority of genetic storage diseases lack efficacious treatments and exhibit a progressive deterioration over time.

Animals can also develop acquired storage diseases when they ingest plants that contain certain inhibitors of lysosomal catabolic enzymes. Over time, these inhibitors can cause the development of acquired neuronal storage disorders. An example of a plant that can cause acquired neuronal storage disorder in animals when consumed is locoweed (Heindel et al., 2017).

Lysosomal storage diseases (LSDs) are a group of disorders that are associated with lipid metabolism. Lysosomal storage disorders (LSDs) are infrequent, inherited conditions resulting from deficiencies in particular enzymes located within cell lysosomes. These abnormalities cause the build-up of substances within lysosomes, resulting in a range of clinical symptoms. LSDs can be classified as either acquired or inherited. Animals afflicted with genetic storage diseases exhibit normal physical appearance upon birth, but subsequently manifest symptoms within the initial weeks or months of their existence. Regrettably, the majority of genetic storage diseases lack efficacious treatments, and their progression tends to deteriorate with time.

Hyperlipidemia is a condition that affects the metabolism of lipids, leading to increased levels of fat in the blood, specifically high levels of cholesterol or triglycerides. Hyperlipidemia in animals can be attributed to several factors, including postprandial hyperlipidemia, medication-induced hyperlipidemia, hereditary hyperlipidemia, and secondary or acquired hyperlipidemia resulting from underlying conditions such as hypothyroidism, Cushing's disease, pancreatitis, and renal disorders.

3.7. ROLE OF LIPIDS IN ANIMAL HEALTH

Lipids are of utmost importance in maintaining optimal animal health. Organic substances are capable of dissolving in organic solvents, while remaining insoluble in water. Lipids consist mainly of carbon and hydrogen, along with small amounts of oxygen. Lipids serve a crucial role in our bodies by functioning as a storage system for energy, much like triglycerides. They possess a significant amount of energy, offering 9 Kcal per gram, in contrast to the mere 4 Kcal per gram found in proteins and carbohydrates. Lipids play crucial roles in the structure of membranes and also have regulatory functions, resembling hormones and fat-soluble vitamins (Braverman & Moser, 2012).

Emulsifiers possess a structure that combines both lipophilic and hydrophilic properties, making them a distinct category of surfactants. One important function is to stabilize the emulsion of lipid and aqueous phases. Emulsifiers play a crucial role in the cattle industry by improving fat optimization. Enhancing the digestion and absorption of fats in animal feed is crucial for fostering growth in pigs and broilers. Emulsifiers play a crucial role in improving the digestibility of various nutrients, including proteins, which ultimately leads to improved zootechnical parameters in animal production.

Chapter 4

Animal Protein Metabolism

Contents

Protein metabolism (Figure 4.1) encompasses the intricate biochemical processes involved in the synthesis of proteins and amino acids (anabolism) as well as the breakdown of proteins into amino acids (catabolism) (Wang et al., 2013). These processes are crucial for the overall health and functionality of animals. The process of protein synthesis involves the translation of genetic information into messenger RNA (mRNA), which is then further translated into proteins. Post-translational modifications further enhance the functional characteristics of these proteins. Proteolysis is a process in catabolism where enzymes break down large proteins into smaller peptides. Subsequently, these peptides undergo enzymatic degradation, resulting in the formation of separate amino acid units. In addition, the liver plays a vital role in synthesizing non-essential amino acids and eliminating ammonia produced during nitrogen metabolism (Wang et al., 2013).

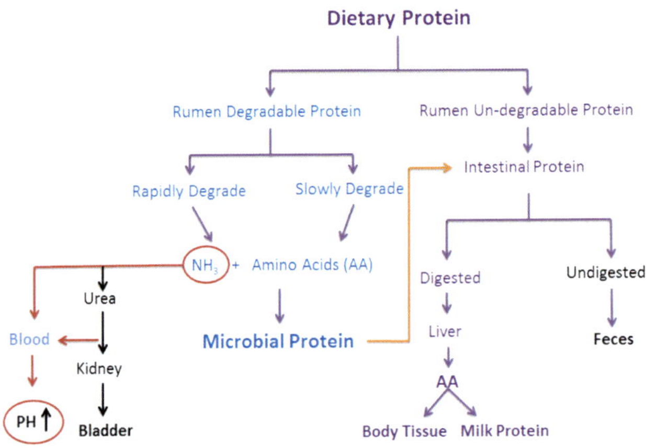

Figure 4.1. *Schematic summary of protein metabolism and digestion in dairy animals.*

Source: Ibtisham et al. (2018).

4.1. PROTEIN STRUCTURE AND FUNCTION

Protein molecules are considerably larger than sugar or salt molecules and consist of multiple amino acids that are connected to form long chains, resembling beads on a string. Proteins consist of approximately 20 naturally occurring amino acids. Proteins that have similar functions exhibit similarity in both their amino acid sequence and composition. The attributes of the amino acids comprising proteins are accountable for the established correlations between structure and function, although it remains impossible to completely elucidate a protein's actions solely based on its amino acid sequence (Yang & Zhang, 2015).

Animal organs typically exhibit a significantly elevated protein content in comparison to blood plasma. As an illustration, the muscle tissue contains roughly 30% protein, the liver contains protein ranging from 20% to 30%, and red blood cells have a protein content of 30%. According to Yang & Zhang (2015), hair, bones, and other organs and tissues with low water content have higher protein percentages. Animals possess a significantly lower amount of free amino acids and peptides compared to protein. Protein molecules are formed within cells through the sequential arrangement of amino acids and are only released into bodily fluids once synthesis is complete.

Some organs contain a significant amount of protein, but the presence of protein in an organism or tissue does not necessarily indicate its importance. In fact, certain crucial proteins, such as hormones and enzymes, are present in very small quantities. The main determinant of the importance of proteins is their function. Thus far, all enzymes that have been identified are proteins. Organisms require proteins, nucleic acids, carbohydrates, and lipids for the processes of synthesis, conversion, and decomposition. Enzymes serve as the catalysts for all metabolic reactions. Life is rendered impossible in the absence of enzymes. Many protein hormones have important regulatory functions. Hemoglobin, an oxygen-carrying protein present in all vertebrates, transports oxygen from the lungs to the body's tissues and organs through the bloodstream. A significant group of structural proteins protects and maintains the integrity of the animal body's structure.

4.1.1. Overview of Protein Structure and Characteristics

** Give primary, secondary, tertiary and quaternary structure of protein

1. **The Protein's Composition of Amino Acids:** Every protein possesses the common feature of being composed of elongated sequences of α-amino (alpha) amino acids. The structural organization of α-amino acids is illustrated. The term "α-amino acids" is used because the α-carbon atom in the molecule possesses both an amino group (–NH_2) and a carboxyl group (—COOH).
2. **Structures of Frequently Encountered Amino Acids:** The amino acids found in proteins exhibit distinct variations in their side chain (R) structures. Glycine is the simplest amino acid, and the letter R represents a hydrogen atom. The letter "R" represents either linear or branched carbon chains in specific amino acids. Alanine is an amino acid, and the letter R represents the methyl group (CH_3). The

alkyl side-chain sequence is finalized by the lengthier R groups of valine, leucine, and isoleucine. The nonpolar alkyl side chains, or R groups, of these amino acids demonstrate their affinity for each other but lack any affinity for water. Although animals have the ability to produce alanine and glycine, plants are capable of synthesizing all other alkyl amino acids. Therefore, valine, leucine, and isoleucine must be obtained through dietary sources (Yang & Zhang, 2015).

Alanine serves as the precursor for the amino acids serine and cysteine, both of which contain three carbon atoms. In serine, the methyl group of alanine is substituted with an alcohol group (CH_2OH), while cysteine contains a mercapto group (CH_2SH). Animals have the ability to synthesize serine, but not cysteine or cystine. The predominant form of cystine in proteins is the oxidized state, characterized by the removal of hydrogen atoms. Cysteine consists of two cysteine molecules linked by a disulfide bond (—S—S—), which forms when a hydrogen atom is removed from each cysteine's mercapto group. Disulfide bonds are essential for protein structure as they facilitate the formation of loops in linear protein chains by linking two separate components of the protein molecule. Proteins contain small amounts of cysteine that have unbound sulfhydryl (—SH) groups (Yang & Zhang, 2015).

Proteins consist of four specific amino acids, namely aspartic acid, asparagine, threonine, and methionine. Every amino acid consists of four carbon atoms. Animals have the ability to synthesize and secrete substantial amounts of asparagine and aspartic acid. Methionine and threonine are essential amino acids that cannot be synthesized by the body and therefore must be obtained from the diet. Methionine is present in minute quantities in most proteins. Additionally, proteins contain proline, which is a secondary amine that forms a ring structure by attaching the amino group (NH_2) to the alkyl side chain. Glutamic acid, an amino acid consisting of five carbon atoms, is also present in proteins (Yang & Zhang, 2015). Glutamic acid and aspartic acid are dicarboxylic acids, meaning they contain two carboxyl groups (COOH).

Similar to asparagine, glutamine is the amide derivative of its corresponding dicarboxylic acid form. In other words, instead of having a carboxyl (COOH) group on its side chain, it has an amide group ($CONH_2$). The majority of proteins contain a high concentration of glutamine and glutamic acid. In plant proteins, these amino acids can constitute over one-third of the total amino acid content. According to Yang & Zhang (2015), animals have the ability to produce both glutamine and glutamic acid.

Tryptophan, phenylalanine, and tyrosine are three additional amino acids that possess an aromatic structure, characterized by the presence of a benzene ring. Animals have the ability to convert phenylalanine into tyrosine, however, they lack the capability to synthesize the benzene ring itself (Yang & Zhang, 2015). Consequently, these three amino acids are essential.

4.2. ANIMAL PROTEIN SYNTHESIS

Protein synthesis refers to the cellular process of creating polypeptide chains from coded combinations of individual amino acids. The protein synthesis process diagram illustrates the presence of a ribosome and mRNA. A protein synthesis diagram illustrates the step-by-step process of DNA transcription to mRNA in the nucleus, followed by mRNA processing and export into the cytoplasm. TRNA molecules possess amino acids that correspond to mRNA codons, facilitating the attachment of ribosomes to mRNA and assisting in the process of translation (Pillai et al., 2007).

Protein synthesis (Figure 4.2) is an essential process in all animals that is required for cell division, growth, and functioning. The process consists of two main stages: translation and transcription. Below is a comprehensive enumeration of the stages involved in protein synthesis:

1. **Transcription:** It initiates in the nucleus of the cell, where messenger RNA (mRNA) is produced by utilizing DNA as a template. To initiate the process of unwinding and separating the DNA strands, the enzyme RNA polymerase binds to the promoter region of a gene on the DNA molecule (Pillai et al., 2007). RNA polymerase synthesizes a complementary mRNA strand by incorporating nucleotides based on the rules of base pairing (A-U and G-C), using one of the DNA strands as a template (Pillai et al., 2007). Once transcription is complete, the mRNA transcript dissociates from the DNA and undergoes processing, which involves the removal of introns (non-coding regions) and the addition of a 5′ cap and poly-A tail (polyadenylation). Subsequently, the fully developed mRNA molecule exits the nucleus and translocates to the cytoplasm.
2. **Translation:** It occurs in the cytoplasm and converts the mRNA sequence into a protein or polypeptide chain. The process initiates when the messenger RNA (mRNA) attaches to a ribosome. Translation involves three essential processes: initiation, elongation, and termination.
   - During initiation, the small ribosomal subunit binds to the mRNA molecule, while the initiator tRNA, which carries the amino acid methionine, binds to the start codon (AUG) on the mRNA. The formation of a functional ribosome occurs when the large ribosomal subunit is joined together.

- During the elongation phase, the ribosome sequentially reads the codons as it moves along the mRNA molecule. The tRNA molecule carries the complementary anticodon and the corresponding amino acid, which forms base pairs with the mRNA codon to establish binding. Every codon corresponds to a distinct amino acid. The ribosome catalyzes the formation of peptide bonds between adjacent amino acids, resulting in the formation of a polypeptide chain.
- Termination occurs when a stop codon (UAA, UAG, or UGA) is identified on the mRNA. At this juncture, the ribosome liberates the fully formed polypeptide chain, signifying the culmination of protein synthesis. Prior to attaining its functional structure, the newly synthesized protein may undergo further modifications and folding (Pillai et al., 2007).

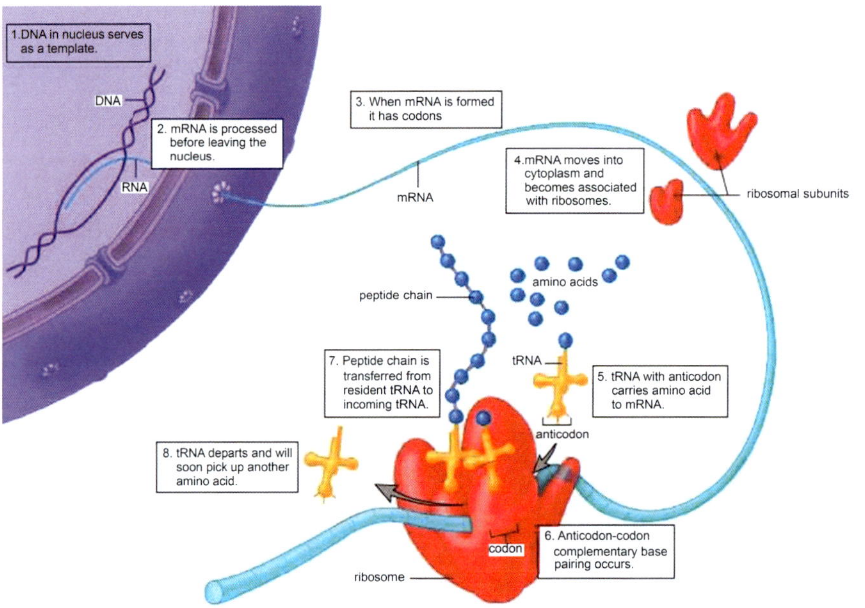

Figure 4.2.*Protein synthesis: Transcription makes RNA, translation builds proteins.*

Source: Nagansurkar (2021).

Protein synthesis is the cellular process responsible for the production of proteins, and ribosomes play a crucial role in facilitating this process. The primary functions they fulfill are as follows:

1. Messenger RNA (mRNA) carries genetic instructions that are decoded by ribosomes. They assemble amino acids in the correct sequence to form a protein chain based on this information.
2. Ribosomes consist of both large and small subunits. Their main role is to facilitate the formation of peptide bonds between amino acids. The proper integration of amino acids into proteins is ensured by this process.
3. In eukaryotic cells, the nucleus provides ribosomes with instructions for protein synthesis. Genes, which are specific sections of DNA, undergo a process called translation to produce messenger RNAs (mRNAs).
4. The ribosome utilizes the mRNA's information to synthesize a protein with a specific sequence of amino acids. Ribosomes are essential components of cellular processes that facilitate cell division, repair, and maintenance in animals.
5. Protein synthesis is crucial as proteins play a vital role in the formation of animal tissues and the functioning of metabolic processes. They serve as the essential building blocks of hormones, enzymes, muscles, and various cellular structures. In order for animals to sustain growth, maintain good health, and ensure productivity, it is essential for them to consume sufficient protein from their diet along with crucial amino acids.
6. Protein synthesis is an essential process that ensures an animal's capacity to grow, maintain good health, and carry out crucial functions (Pillai et al., 2007).

4.3. ANIMAL PROTEIN DEGRADATION

The quantities of proteins in cells are influenced by both the rates of synthesis and degradation. The differential rates of protein degradation play a vital role in cellular regulation, as the lifespans of proteins in cells can vary significantly, spanning from minutes to several days.

A plethora of rapidly degraded proteins function as transcription factors and other regulatory molecules. In order to quickly adjust their levels in response to external stimuli, these proteins must undergo rapid turnover (Santos et al., 2016). Additionally, there exists an alternative method to regulate intracellular enzyme activity: specific signals induce rapid degradation of other proteins. Moreover, cells promptly detect and degrade faulty or impaired proteins, thereby minimizing the impact of mistakes that occur during the production of newly synthesized proteins. In eukaryotic cells, protein breakdown is facilitated by two primary pathways: lysosomal proteolysis and the ubiquitin-proteasome system (Santos et al., 2016).

4.3.1. The Ubiquitin-Proteasome Pathway

Ubiquitin functions as a signal for the main process of specific protein breakdown in eukaryotic cells. This process specifically targets nuclear and cytosolic proteins for fast degradation. The polypeptide ubiquitin consists of 76 amino acids and is conserved in all eukaryotes, such as plants, mammals, and yeasts. Ubiquitin binding to the amino group of a lysine residue serves as a signal for protein degradation. Subsequent ubiquitin's are added to form a multiubiquitin chain. The proteasome, a complex enzyme composed of multiple subunits, has the ability to recognize and degrade polyubiquitinated proteins. Throughout the process, ubiquitin is liberated, enabling its reuse in subsequent cycles. Energy is required for both the attachment of ubiquitin and the breakdown of marked proteins.

Proteins are marked for rapid degradation by the covalent attachment of multiple ubiquitin molecules. E1 is the enzyme responsible for the initial activation of ubiquitin. One of the many ubiquitin-conjugating enzymes (E2) receives the ubiquitin that has been activated afterwards. Subsequently, the ubiquitin molecule is frequently conveyed to an E3 ubiquitin ligase prior to being transferred to a specific target protein. The polyubiquitinated proteins are broken down by the proteasome, a protease complex, after the addition of multiple ubiquitins (Santos et al., 2016).

The stability of numerous proteins is determined by their capacity to undergo ubiquitination, which serves as a signal for rapid degradation of the proteins. Ubiquitination entails a series of distinct steps. Initially, ubiquitin is rendered active through its interaction with the enzyme E1, thereby triggering ubiquitin activation. Afterwards, the ubiquitin is transferred to ubiquitin-conjugating enzyme (E2), which is another enzyme. Ubiquitin ligase, also referred to as E3, is responsible for specifically identifying the appropriate substrate proteins and facilitating the ultimate transfer of ubiquitin to the target protein. Some specific scenarios require the movement of ubiquitin from E2 to E3, which is then attached to the target protein. Alternatively, ubiquitin can associate with E3 to create a complex and then be directly transferred from E2 to the target protein. Although the majority of cells possess numerous E2s and multiple families of E3 enzymes, they only possess a single E1 enzyme. The ubiquitin-proteasome (Figure 4.3) system selectively directs the degradation of cellular proteins through the action of enzymes from the E2 and E3 families, which recognize specific substrate proteins.

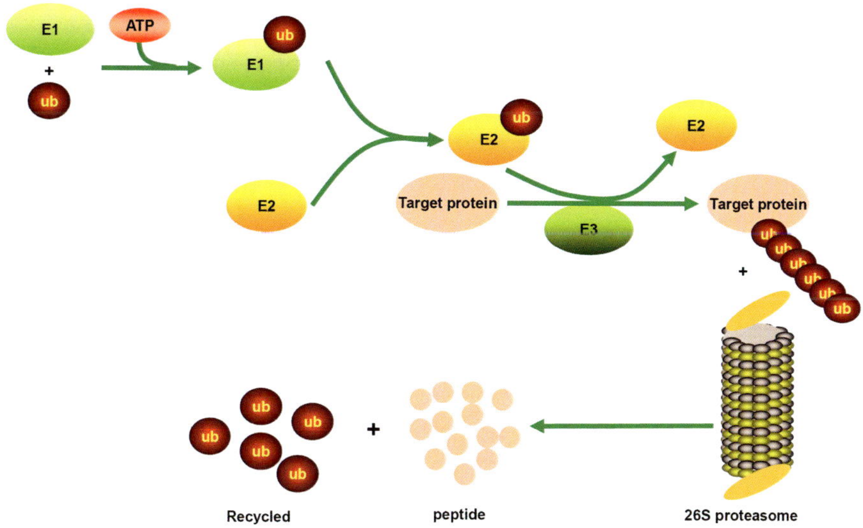

Figure 4.3. *The ubiquitin proteasome pathway.*

Source: Tu et al. (2012).

Ubiquitination and proteolysis, which are regulated processes, selectively target various proteins involved in crucial physiological functions such as gene expression and cell division. The proteins known as cyclins, which regulate the progression of an animal cell through its division cycle, provide a fascinating example of this type of controlled degradation. Cyclin B, a regulatory subunit of the protein kinase Cdc2, plays a crucial role in controlling the initiation of mitosis in all eukaryotic cells. The binding of Cyclin B to Cdc2 is essential for the activation of Cdc2 kinase, which phosphorylates various cellular proteins to initiate mitotic processes, including chromosomal condensation and disruption of the nuclear envelope. During the final stages of mitosis, cyclin B is broken down through a process called proteolysis, which is facilitated by ubiquitin. This degradation is also initiated by Cdc2 (Santos et al., 2016). When cyclin B is degraded, Cdc2 becomes inactive, allowing the cell to transition from mitosis to the interphase of the next cell cycle. The destruction box, a 9-amino-acid region of cyclin B, specifically directs the process of ubiquitination of cyclin B. The mutations in this sequence prevent the proteolysis of cyclin B and hinder the progression of dividing cells in mitosis. This emphasizes the importance of regulated protein degradation in controlling the crucial process of cell division (Santos et al., 2016).

4.3.2. Lysosomal Proteolysis

The process of lysosomal protein absorption is one of the primary mechanisms through which proteins are degraded in eukaryotic cells. Lysosomes, which are membrane-bound organelles, contain a diverse range of digestive enzymes, including multiple proteases. They perform various roles in cell metabolism, including the gradual replacement of cytoplasmic organelles and cytosolic proteins, as well as the breakdown of extracellular proteins absorbed through endocytosis (Santos et al., 2016).

The lysosome's capacity to sequester proteases and other hydrolytic enzymes prevents the intracellular components from undergoing uncontrolled degradation. Consequently, lysosomal proteolysis (Figure 4.4) requires the prior uptake of cellular proteins by lysosomes for degradation. One mechanism for the uptake of cellular proteins is the production of vesicles, also known as autophagosomes. These vesicles are formed from the endoplasmic reticulum and encase small portions of cytoplasm or cytoplasmic organelles within membranes. After merging with lysosomes, the substances inside these vesicles are decomposed by the degradative lysosomal enzymes. Cytoplasmic proteins with long lifespans undergo gradual degradation due to the indiscriminate uptake of proteins into autophagosomes.

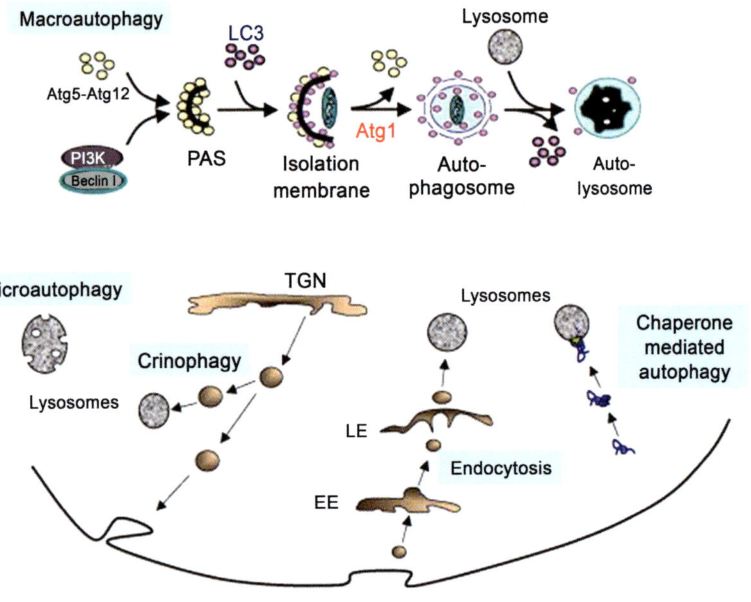

Figure 4.4.*Lysosomal proteolysis.*

Source: Bechet et al. (2005).

Proteases are a type of hydrolytic enzymes that are present in lysosomes and are responsible for the breakdown of proteins. Lysosomes can absorb proteins from cells by merging with autophagosomes, which are formed when parts of the endoplasmic reticulum enclose areas of cytoplasm or organelles such as mitochondria (Santos et al., 2016). This union results in the production of a phagolysosome, which is responsible for the breakdown of the contents of the autophagosome.

However, lysosomes do not indiscriminately absorb all proteins. For example, when deprived of cellular resources, lysosomes possess the capacity to selectively internalize and degrade specific proteins found in the cytosol. In such situations, the amino acid sequences of the proteins that are broken down by lysosomal proteases bear a strong resemblance to the widely accepted sequence Lys-Phe-Glu-Arg-Gln, which is likely responsible for directing these proteins to lysosomes. The degradation of these proteins in the lysosome also necessitates the presence of a molecular chaperone from the Hsp70 family. This chaperone is likely responsible for unfolding the polypeptide chains as they traverse the lysosomal membrane. It is hypothesized that the proteins susceptible to degradation via this pathway are generally durable yet expendable proteins. Proteins are sacrificed during periods of famine to generate energy and amino acids, facilitating crucial metabolic processes (Santos et al., 2016).

4.4. ANIMAL AMINO ACID METABOLISM

The main sources of amino acids for the human body include the proteins obtained from animal diets, the non-essential amino acids synthesized by the liver, and the constantly metabolized and reassembled amino acids derived from the body's protein stores (Wu et al., 2014). Approximately 10% of the peptide bonds undergo hydrolysis due to the activity of gastric juice in the stomach, which initiates protein digestion. The composition of gastric juice consists primarily of water, inorganic ions, hydrochloric acid, various enzymes, and other proteins, accounting for over 99%. The pain associated with a gastric ulcer is primarily caused by the irritated tissue resulting from the acidic stomach fluid. Figure 4.5 **shows mechanism of** metabolic pathways of amino acids for ketosis in dairy cows. Gastric juice contains hydrochloric acid (HCl), which is secreted by the glands in the stomach lining. The pH of freshly released gastric juice is approximately 1.0, but the contents of the stomach can increase it to 2.5 or higher. HCl can denature food proteins by causing the unfolding of protein molecules, thereby facilitating more efficient enzyme activity along their chains. Pepsinogen, an enzymatically inactive precursor produced by the cells in the stomach wall, serves as the primary component responsible for digestion in gastric juice.

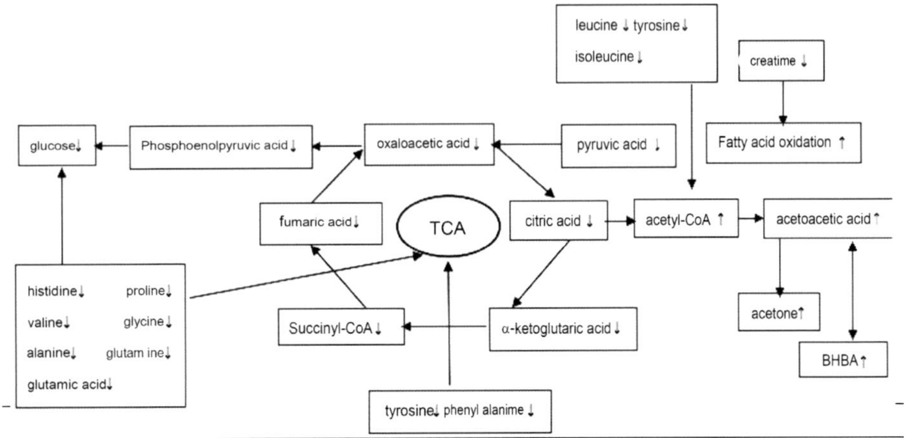

Figure 4.5.*Metabolic pathways of amino acids for ketosis in dairy cows.*

Source: Wang et al. (2016).

Following a period of abstaining from food, a decrease in pH initiates a sequence of chemical processes in the stomach that transform pepsinogen into its active state, known as pepsin. Pepsin catalyzes the hydrolysis of peptide bonds in protein molecules. It primarily acts on links that contain aromatic amino acids such as tryptophan, tyrosine, and phenylalanine, as well as methionine and leucine. However, it also has a relatively wide range of targets. The small intestine is the site where the process of protein digestion is completed (Wu et al., 2014).

After the hydrolysis of dietary proteins, the resulting free amino acids are incorporated into the pool of amino acids that can now be used in metabolic processes. This pool comprises the non-essential amino acids synthesized in the liver, along with the amino acids that are reclaimed from the body's own proteins (Wu et al., 2014). Most of the amino acid pool is used for synthesizing proteins and other nitrogen-containing molecules, such as hormones, neurotransmitters, DNA bases, and others. Animals have the ability to utilize amino acids as a source of energy in certain metabolic circumstances. When the amino acids present in the pool are not required for biological processes, they undergo degradation and release nitrogen into the urine (Wu et al., 2014).

This intricate process ensures the utilization, production, and reutilization of amino acids, which is essential for protein synthesis, energy generation, and overall well-being.

1. **Nutritional Consumption:**Animals acquire amino acids predominantly from their diet. Amino acids can be categorized into two groups: essential amino acids and nonessential amino acids (Wu et al., 2014).

- **Essential Amino Acids:** The body does not have the ability to synthesize any essential amino acids. Consequently, they must come from food sources. The nine essential amino acids are histidine, isoleucine, leucine, lysine, methionine, phenylalanine, threonine, tryptophan, and valine. These are the amino acids that the body cannot produce on its own and must be obtained through diet. The term "non-essential amino acids" refers to the amino acids that the body can synthesize on its own.
- **Non-Essential Amino Acids:** When an amino acid is synthesized by animal organisms without requiring external sources from food, it is referred to as non-essential. Alanine, arginine, asparagine, aspartic acid, cysteine, glutamic acid, glutamine, glycine, proline, serine, and tyrosine are classified as non-essential amino acids.

Proteins can be classified as complete or incomplete depending on the arrangement of amino acids. Complete proteins contain all nine essential amino acids in adequate proportions. Complete proteins encompass the proteins present in meat, fish, milk, eggs, and other animal-derived products. Inadequate concentrations of one or more essential amino acids result in incomplete proteins (Wu et al., 2014).

2. **Digestion and Absorption:** Proteolytic enzymes present in the stomach and small intestine degrade food proteins into smaller peptides and amino acids. The intestinal epithelium facilitates the absorption of these amino acids into the bloodstream.
3. **Transportation and Distribution:**The circulatory system transports amino acids to various tissues. The liver plays a crucial role in regulating the levels of amino acids in the body. Certain amino acids are employed in protein synthesis, while others serve as foundational components for other compounds, such as hormones and neurotransmitters.
4. **Synthesis of Proteins:**During the process of translation, amino acids are assembled into polypeptide chains, as previously stated. Ribosomes synthesize polypeptide chains by decoding the genetic instructions encoded in mRNA. The polypeptides subsequently undergo folding to form functional proteins.
5. **Catabolism and Deamination:** This refer to the breakdown of molecules and the removal of amino groups, respectively. Excess amino acids are catabolized in order to generate energy. Deamination

is the act of eliminating the amino group (NH) from the amino acid. The presence of the amino group results in the formation of the toxic compound ammonia (NH). The liver converts ammonia into urea, which is subsequently excreted by the kidneys.

6. **Transamination:** It is a biochemical process that facilitates the conversion of one amino acid to a different amino acid. To generate a novel amino acid, the amino group of one amino acid is transferred to a keto acid.
7. **Conversion of Amino Acids:** Amino acids have the ability to undergo conversion from one type to another. Glutamine can be synthesized from glutamate, for example. Asparagine can be synthesized from aspartate.
8. **Special Pathways:** Specific amino acids participate in specific processes; for instance, arginine is involved in the urea cycle. Serotonin is a neurotransmitter synthesized from tryptophan. Catecholamines, such as adrenaline and dopamine, serve as precursors to tyrosine.
9. **Nitrogen Balance:** The nitrogen intake from amino acids and the nitrogen excretion as urea in animals are in equilibrium. A positive nitrogen balance occurs when an organism experiences growth or healing. A negative nitrogen balance occurs when an individual is in poor health or malnourished.

Ultimately, the metabolism of amino acids is a constantly changing process that encompasses the breakdown, synthesis, digestion, and interconversion of these molecules. The statement ensures that animals receive the necessary constituents for their growth, upkeep, and overall well-being (Wu et al., 2014).

4.5. PROTEIN METABOLISM REGULATION

Protein accretion takes place when the rate of protein synthesis is higher than the rate of proteolysis, while protein loss occurs when the opposite happens. Proteins in tissues undergo constant synthesis and degradation. The equilibrium between these two processes governs the quantity of protein that exists in the body. Insulin plays a significant role in regulating protein synthesis and breakdown through physiological, genetic, and environmental mechanisms. Additional variables that impact this regulation encompass nutrition, temperature, and hormones.

In a healthy animal with a stable body mass, protein synthesis, protein breakdown, and protein intake are all in equilibrium. Insulin typically stimulates protein synthesis and maintains protein balance. However, other hormones play a vital role, particularly in skeletal muscle, which serves as the body's primary protein storage. These hormones include thyroxine, glucocorticoids, growth hormone, and insulin-like growth factor-l (IGF-l). Amino acids stimulate the secretion of insulin; however, excessive protein intake does not lead to retention and does not contribute to an increase in lean body mass. Growth hormone, IGF-1, anabolic steroids, and exercise are external factors that stimulate an increase in lean body mass (Morrison & Laeger, 2015). β2-adrenergic stimulants exhibit anabolic effects in animals, but there is presently inadequate evidence to substantiate this effect in humans.

The first step in the catabolism of an amino acid involves the elimination of the NH_2 group and its substitution (deamination) with a C=O group, resulting in the formation of an oxo or keto acid. According to Morrison & Laeger (2015), multiple oxo acids can undergo oxidation to yield carbon dioxide and water. These substances are metabolic intermediates that are either directly involved in or closely related to the metabolic pathways that have been previously mentioned, such as glycolysis and the TCA cycle.

Out of the amino acids known as glucogenic, 13 can only be converted into glucose through a process called gluconeogenesis. One amino acid can only be converted into fat through a process called ketogenic synthesis. The remaining 13 amino acids can be converted into either glucose or fat (Morrison & Laeger, 2015).

The conversion of amino acids into glucose becomes essential in the absence of glucose intake, particularly for ensuring a steady supply of glucose to the central nervous system (CNS). Both the kidney and the liver are involved in the process of glucose biosynthesis. Ammonia is generated through the elimination of the NH_2 group from an amino acid, and it possesses toxic properties. Nevertheless, once it is created, it is rapidly broken down and the concentration of ammonia in the bloodstream is generally very low (Morrison & Laeger, 2015).

The liver utilizes the urea cycle to convert ammonia and oxygen into urea. Nevertheless, in peripheral tissues, NH_2 groups are transferred to keto acids, particularly α-ketoglutarate, which serves as an intermediate in the tricarboxylic acid cycle (TCA cycle), leading to the production of glutamate (containing one NH_2 group) and glutamine (containing two NH_2 groups). In addition, they are converted into pyruvate, which is the corresponding keto acid to alanine. Alanine and glutamine are the primary means by which NH_2 groups are transported from the peripheral areas to the liver and kidney. Glutamate serves as both a precursor for renal gluconeogenesis and a buffer against excessive acid secretion. Alanine is the main precursor of hepatic gluconeogenesis (Morrison & Laeger, 2015).

4.5.1. The Urea Cycle

** Give fig. of Urea Cycle

The liver synthesizes urea through a cyclic mechanism. Ornithine is the fundamental compound used to build the urea molecule, while arginine and citrulline are involved as intermediate compounds in this process. Ammonia is released from amino acids, specifically glutamine and alanine, and then combined with ATP and carbon dioxide to create carbamoyl phosphate in the mitochondrion of hepatocytes. The cytosol is where urea is synthesized and subsequently excreted. Glucagon, in addition to its role in regulating the liver's uptake of amino acids, activates the entire process.

4.5.2. Gluconeogenesis

During periods of food restriction, such as fasting and undernutrition that worsens sepsis and trauma, the body sustains its requirement for glucose by synthesizing it from smaller molecules like glycerol, lactate, and amino acids. Essentially, gluconeogenesis is the opposite of glycolysis, but it bypasses the three irreversible steps of glycolysis. Most amino acids undergo deamination and enter the tricarboxylic acid (TCA) cycle through various pathways. Pyruvate is the main precursor and direct oxo acid counterpart of alanine, and it is closely linked to lactate. The process involves carboxylation to produce oxaloacetate, which is a constituent of the TCA cycle. Subsequently, phosphoenolpyruvate carboxykinase (PEPCK) converts it into phosphoenolpyruvate (Morrison & Laeger, 2015).

The second phase of the glycolytic process, which is capable of being reversed to generate glucose, involves the presence of phosphoenolpyruvate. Fructokinase and glucose-6-phosphatase prevent the irreversible steps of glycolysis. Fructose-6-phosphate is transformed into fructose-1,6-bisphosphate by phosphofructokinase, while glucose is produced from glucose-6-phosphatase. Glycerol, the third main precursor, undergoes a transformation into dihydroxyacetone and glycerol-3-phosphate. These compounds are integral components of the second phase of the glycolytic pathway (Morrison & Laeger, 2015).

Gluconeogenesis is regulated by the rate at which precursors are transported to the liver from peripheral tissue, as well as the activity of glucose-6-phosphatase, fructose 1,6 bisphosphatase, and PEPCK. Peripheral tissue events, such as the production of lactate from muscle during exercise or shock, glycerol from lipolysis during starvation or after adrenergic stimulation, and amino acids during injury or starvation due to the inhibition of insulin release or resistance to its action, control the supply of gluconeogenic precursors to the liver. Glucagon is the primary hormone responsible for activating various enzymes, although other hormones such as cortisol and low insulin also exert influence.

4.6. PROTEIN METABOLISM DISORDERS IN ANIMALS

Metabolic illnesses resulting from production are primarily caused by an imbalanced consumption of a specific nutrient. Occasionally, an animal's dietary intake of a nutrient may decrease significantly, even though its metabolic demand for that nutrient remains high. Some examples include pregnancy toxemia in ewes, protein-energy deficiency in beef cattle, dairy cow syndrome, and hyperlipemia in ponies. Moreover, certain diseases may exacerbate if producers are compelled to exclude nutrients from animals that already have insufficient diets, primarily due to financial concerns (Mizushima & Klionsky, 2007).

Below are several instances of protein metabolism disorders observed in animals:

1. **Gestation Toxemia in Ewes:** This refers to a condition characterized by the presence of toxins in the bloodstream of female sheep.
   - Occurs during the early stages of lactation or towards the end of pregnancy.
   - Insufficient energy intake leads to the breakdown of excessive protein and body fat.
   - Ewes experience weakness, lethargy, and possibly neurological symptoms.
   - This condition is commonly seen in ewes with multiple lambs.
2. **Protein-Energy Malnutrition in Beef Cattle:** The insufficient intake of calories and protein from food is the cause. Mizushima & Klionsky (2007) reported that low milk production, weight loss, and poor growth are the consequences of this condition. Calves may experience weakness, subpar coat quality, and muscle atrophy.
3. **Obesity in Dairy Cattle:**
   - Acetonemia or ketosis is another term for this condition.
   - It happens when milk production exceeds the rate at which it is being consumed.
   - Results in the excessive mobilization of fat and the accumulation of ketone bodies. The affected cows display symptoms of weakness and loss of appetite, and the presence of acetone in their breath may result in a pleasant odor.

4. **Hyperlipemia in Ponies:** Hypertriglyceridemia is a metabolic condition characterized by elevated levels of triglycerides in the bloodstream. It is commonly seen in horses that are obese or overweight.

The user's text consists of a bullet point symbol. Hyperlipemia can be caused by inadequate feed intake, illness, or stress. Clinical signs of hyperlipemia include lethargy, anorexia, and the presence of yellowish fat deposits in the blood arteries.

To mitigate the adverse impact of these illnesses on animal health, it is crucial to promptly identify them and provide efficient treatment.

Diseases related to the metabolism of animal proteins are diagnosed through a combination of laboratory testing, clinical assessment, and specialized diagnostic instruments. Common methods include:

- **Presentation of Symptoms and Medical Background:**Initial clues can be identified through the observation of clinical indicators. Indicators such as weakness, weight loss, a low-quality coat, or specific symptoms related to the disorder may be present. A comprehensive history provided by the owner or caregiver helps in identifying the need for dietary modifications, stress management strategies, or recent alterations.
- **Laboratory Tests:**
  - **Hematological Analyzes:** Various metabolites, such as ketone bodies and amino acid levels, can be utilized to quantify deviations from normality.
  - **Urinalysis:** Identifying specific amino acids or metabolites in urine can yield valuable insights.
  - **Analysis of Blood Serum Components:** Obtaining measurements of glucose, electrolytes, and liver enzymes provides valuable information.
- **Evaluation of Body Condition:**Examining the animal's physical condition assists in assessing its overall well-being and nutritional status. Fluctuations in physical condition may indicate the presence of different metabolic disorders or malnutrition related to protein and energy deficiencies.
- **Measurement of Fat/Protein Levels Proportions:**Imbalances can be identified by comparing the quantities of fat and protein in milk and blood. An example of this is when there are higher proportions of fat compared to protein, which could suggest the presence of metabolic issues.

- **Amino Acid Disorder-Specific Tests:** Concentrations of amino acids in bodily fluids can be utilized to detect disorders that impact the metabolism of amino acids, such as phenylketonuria and tyrosinemia.
- **Monitoring Mineral Levels:** Monitoring mineral levels, such as calcium, magnesium, and phosphorus, is crucial for conditions like post-parturient hemoglobinuria or milk fever.
- **Predictive Examinations:** Certain diagnostic procedures have the ability to forecast the occurrence of a disease prior to the manifestation of any symptoms. Early detection allows for prompt intervention. It is important to remember that effectively managing and treating issues related to the metabolism of animal protein requires a precise and prompt diagnosis (Mizushima & Klionsky, 2007).

4.7. THE SIGNIFICANCE OF PROTEINS IN MAINTAINING ANIMAL HEALTH

Proteins are the fundamental components of animal tissues, including skin, muscles, wool, feathers, tendons, and eggs. These proteins are crucial for sustaining life. Moreover, proteins play crucial roles in various biological processes such as transportation (lipoproteins), immune responses (immunoglobulins), metabolic reactions (enzymes), and regulatory functions (hormones). Proteins can provide energy when necessary. Proteins play a crucial role in the formation of animal tissue structures, such as muscle, as well as in the functioning of metabolic processes, including enzymes and hormones. Therefore, the sustenance of vital biological functions relies on the process of protein synthesis. Adequate intake of protein and amino acids is essential for promoting growth, maintaining health, and maximizing productivity in animals that are bred for food production. Ruminant animals possess the ability for their intestinal microbiota to produce proteins from sources that are not proteins themselves (Cray et al., 2009).

The protein requirement varies based on the life stage, with higher amounts needed during periods of rapid growth and during pregnancy and lactation in young animals. Proteins consist of hydrogen, oxygen, and carbon, similar to other macronutrients. Moreover, proteins contain sulfur and nitrogen, which are present in certain amino acids. Proteins are unique among other foods in animal nutrition due to their nitrogen content, which makes them highly distinguishable in terms of their ability to be absorbed, metabolized, and eliminated by the animal.

4.7.1. The Process of Increasing, Restoring, and Substituting Damaged or Deteriorated Bodily Tissues

Protein plays a crucial role in maintaining animal health, particularly in terms of tissue growth, repair, and replacement. Protein provides the essential components necessary for the growth of tissues, division of cells, and development of organs during periods of growth. Furthermore, protein plays a crucial role in the process of rejuvenating and restoring damaged tissues. Protein aids in the repair of muscles by creating new muscle cells through synthesis following physical activity. Moreover, protein plays a crucial role in maintaining healthy skin, facilitating normal bone remodeling, and strengthening the immune system. Protein turnover ensures the replacement of old proteins with new ones, thereby maintaining the overall health of tissues. In order to perform these essential functions, it is necessary to consume a well-rounded diet that contains a sufficient amount of protein (Cray et al., 2009).

4.7.2. Antibiotic Production for Animal Disease Prevention

Proteins play a crucial role in maintaining the health of animals, particularly in the production of antibiotics that protect them from diseases. Probiotics are living microorganisms that are beneficial to health and mainly comprise of yeast and beneficial bacteria. In addition, they have the ability to synthesize antimicrobial compounds such as bacteriocins or proteins. These compounds enhance the immune system, regulate the microbiota in the gut, and inhibit the formation of harmful microorganisms. Antimicrobial peptides (AMPs) possess a wide range of antibacterial effects, inhibiting the acquisition of resistance genes by bacteria. This is in contrast to conventional antibiotics, which may only target specific types of bacteria (Cray et al., 2009). Probiotics offer a beneficial alternative to antibiotics, ensuring the safety of food while avoiding any adverse impacts on animals and promoting overall well-being. The objective of this research is to eliminate any gaps in knowledge and demonstrate that probiotic supplements can effectively replace antibiotics.

4.7.3. Synthesis of Digestive Enzymes for the Purpose of Breaking Down Food Particles

Proteins play a crucial role in maintaining the well-being of animals, particularly in the production of digestive enzymes. Enzymes, which are biologically active proteins, aid in the process of digestion by breaking down nutrients into smaller molecules. Enzymes enhance the digestion of proteins, carbohydrates, and lipids in animal feed, thereby optimizing nutrient absorption in animals raised for food production. Enzymes such as lipase, protease, amylase, and others ensure efficient nutrient absorption in the gastrointestinal system. Enzymes, which are proteins, are specifically designed for their role in digestion. Ensuring the well-being of animals and optimizing the availability of nutrients requires maintaining the stability of feed during digestion and passage through the gastrointestinal tract. Amino acids, which are the fundamental constituents of proteins, are assimilated and utilized for the purposes of growth, repair, and various other physiological processes, provided that proteins are adequately digested. Proteins play a crucial role in the well-being and productivity of animals by facilitating enzymatic digestion (Cray et al., 2009).

Proteins play a crucial role in the synthesis of hormones, serving as the essential components for these important chemical messengers. Amino acids obtained from protein consumption are utilized by the body to synthesize hormones. Inadequate protein intake would hinder the production of hormones responsible for regulating growth, mood, and metabolism. Protein plays a crucial role in maintaining an animal's overall health by promoting hormonal balance (Cray et al., 2009).

4.7.4. The Manufacturing of Products Such as Meat, Milk, Eggs, and Wool

Proteins play a crucial role in maintaining the health of animals, particularly in the production of essential resources such as meat, milk, eggs, and wool. The growth and maintenance of meat, specifically muscle tissue, rely on proteins, ensuring the production of meat of superior quality. Casein and whey proteins are utilized in the production of milk, which has an impact on its composition and yield. Egg whites, also known as albumen, facilitate the growth of the embryo within the eggs due to the presence of proteins. Moreover, the fibrous protein keratin is responsible for the formation of the structural foundation of wool fibers, thereby affecting the growth and quality of wool. The abundance and excellence of these invaluable animal-derived products are directly influenced by adequate protein nutrition, which supports sustainable agriculture and the overall welfare of animals (Cray et al., 2009).

Chapter 5

Enzymes in Animal Biochemistry

Contents

Enzymes play a crucial role in animal biochemistry as biological catalysts. These exceptional proteins lower the activation energy required for specific processes, thereby accelerating chemical reactions (Aehle, 2007). Enzymes play a vital role in facilitating the metabolic processes that enable living systems to extract energy from their surroundings. An instance of this is the process of oxidizing sucrose, which results in the production of carbon dioxide (CO_2) and water (H_2O). This reaction yields a substantial quantity of energy that animals utilize for purposes such as locomotion, sensation, and cognition. However, in the absence of enzymes, these specific types of chemical reactions would not occur at a sufficiently rapid rate to sustain life. Enzymes exhibit remarkable selectivity and function effectively in aqueous solutions under mild pH and temperature conditions. They strategize numerous consecutive reactions that break down nutrition molecules, convert and retain chemical energy, and construct substantial biological macromolecules from more basic components.

Enzyme evaluations conducted on tissues, plasma, and erythrocytes are particularly useful in diagnosing certain illnesses and deficiencies, especially those of a hereditary nature. Enzymes play a crucial role in various fields such as food processing, medicine, the chemical industry, and agriculture, as well as being used as diagnostic tools (Aehle, 2007). Understanding the organization, function, and classification of enzymes is crucial for comprehending the intricate biochemical processes that contribute to the well-being and care of animals. This chapter will discuss different facets of enzymes in animal biochemistry, including the structure and function of enzymes, enzyme kinetics in animals, enzyme regulation in animals, enzyme inhibition in animals, enzyme assays in animals, disorders of enzyme function in animals, and the role of enzymes in animal health.

5.1. THE COMPOSITION AND ROLE OF ENZYMES

Enzymes are biological polymers that facilitate biochemical reactions. Enzymes, predominantly composed of proteins, possess catalytic properties that are indispensable for performing diverse functions. Enzymes play a vital role in sustaining life by facilitating numerous chemical reactions and metabolic processes within cells. The enzymes responsible for initiating the initial phase of the metabolic process interact with a specific substance referred to as the substrate (Copeland, 2023). Enzymes convert substrates into distinct molecules known as products.

Enzymes are crucial for the functioning of biological processes, and their regulation has had a significant impact on clinical diagnosis. Enzymes consist primarily of proteins, except for ribozymes, which are a type of RNA catalysts.

The term "ribozyme" is derived from the ribonucleic acid enzyme. Ribozymes are RNA molecules that catalyze processes within or between other RNAs. Enzymes are present in every tissue and fluid of animals. Enzymes present within cells facilitate all reactions that take place in metabolic pathways. Enzymes within the circulatory system regulate the process of blood clotting, while enzymes within the plasma membrane govern the catalytic reactions that take place within cells in response to cellular signals (Copeland, 2023).

5.1.1. Structure of Enzymes

The three-dimensional structure of an enzyme (Figure 5.1)is determined by the linear chain of amino acids that it is composed of. The arrangement of the enzyme is dictated by the sequence of amino acids, which also signifies the enzyme's ability to facilitate chemical reactions. Elevated temperatures induce the denaturation of the enzyme's structure, resulting in a reduction in its activity. This decrease in activity is typically influenced by temperature.

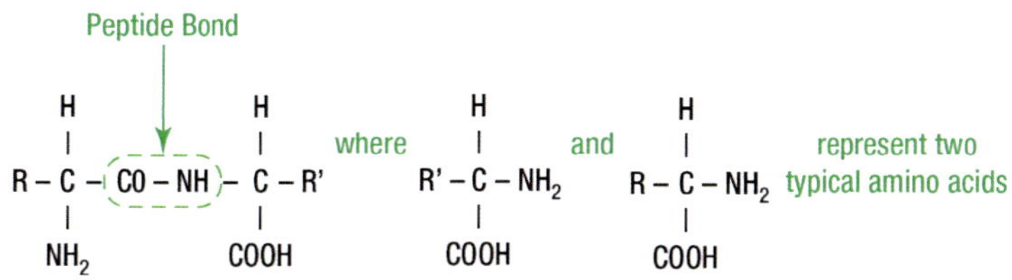

Figure 5.1. *Enzymes are proteins made of amino acids linked by bonds.*

Source: Worthington Biochemical (2024).

Enzymes typically have a larger size compared to their substrates. For instance, fatty acid synthase consists of an average of 2,500 residues, whereas some other enzymes may have as few as 62 amino acid residues. Only a small fraction of the structure is involved in catalysis and is positioned next to the binding sites. The active site of the enzyme comprises both the binding site and the catalytic site. Several ribozymes serve as RNA-based biological catalysts (Copeland, 2023).

Historically, the nomenclature of enzymes was assigned by the individual who made the initial discovery. Increased research efforts resulted in a more comprehensive categorization.

The International Union of Biochemists (IUB) classifies enzymes into six functional categories based on the specific type of reaction they catalyze (Figure 5.2).

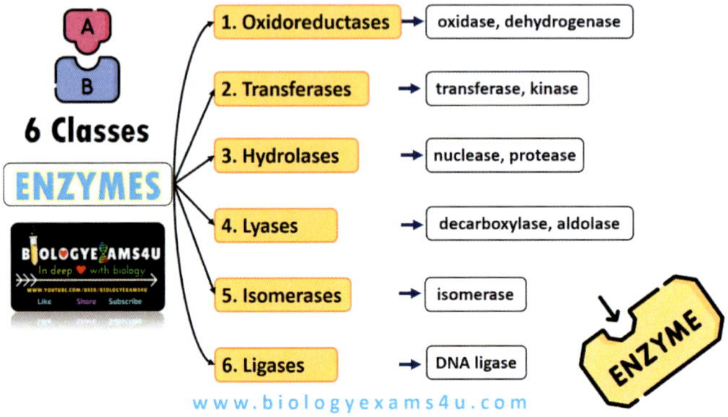

Figure 5.2. *A comprehensive classification of the enzymes.*

Source: Haneef (2023).

- Oxidoreductases are enzymes that catalyze the oxidation reaction, where electrons are transferred from one molecule to another. Transferases are a type of enzymes that facilitate the transfer of functional groups from one molecule to another, known as the donor and acceptor molecules, respectively (Copeland, 2023).
- Hydrolases are a type of enzymatic proteins that facilitate the hydrolysis reaction by adding water to break and hydrolyze chemical bonds.
- Lyases catalyze the formation of double bonds by either adding water, carbon dioxide, or ammonia across them, or by removing these substances.
- Isomerases are enzymes that catalyze the structural changes in molecules, leading to a modification of the molecule's form.
- Ligases are enzymes that catalyze the ligation process.

5.1.2. Cofactors

Enzymes can bind to non-protein molecules known as cofactors. An enzyme necessitates the presence of a cofactor in order to function. An apoenzyme refers to the protein portion of an enzyme that requires a cofactor to function. A holoenzyme is composed of an enzyme and its cofactor (Copeland, 2023). Enzymes possess three distinct types of cofactors: Prosthetic groups are cofactors that form a permanent bond with an enzyme. FAD, or flavin adenine dinucleotide, is a prosthetic group found in many enzymes. An enzyme only binds to a coenzyme during catalysis. It dissociates from the enzyme during all other periods. A commonly occurring coenzyme is NAD. Metal ions are required for catalysis in certain enzymes to form coordinating bonds at the active site. Several enzymes utilize a metal ion cofactor (Copeland, 2023).

5.1.3. Enzyme Functions

1. **Oxidoreductases:**These enzymes, such as pyruvate dehydrogenase, facilitate the conversion of pyruvate into acetyl coenzyme A, as well as other enzymes that facilitate oxidation and reduction reactions.
2. **Transferases:** Transaminases, also known as aminotransferases, are enzymes that catalyze the transfer of amino groups from one molecule to another (Copeland, 2023).
3. **Hydrolases:** They promote the hydrolysis of a bond. The enzyme pepsin hydrolyzes protein peptide linkages.
4. **Lyases:** These are enzymes that catalyze the breaking of chemical bonds. These substances disrupt chemical bonds without requiring the presence of a catalyst. A specific instance is when the compound fructose-1, 6-bisphosphate is divided into two smaller compounds, glyceraldehyde-3-phosphate and dihydroxyacetone phosphate, by the enzyme aldolase. This process occurs within the metabolic pathway known as glycolysis.
5. **Isomerases:** These are enzymes that catalyze the conversion of one isomer to another. They facilitate the isomerization of the compound through catalysis. For example, in the process of glycogenolysis, where glycogen is converted into glucose to release energy quickly, phosphoglucomutase facilitates the conversion of glucose-1-phosphate into glucose-6-phosphate by shifting a phosphate group from one location to another within the same substance.
6. **Ligases:**It facilitate the binding of two molecules. For example, DNA ligase catalyzes the formation of a phosphodiester bond to facilitate the fusion of two DNA fragments.

5.2. ANIMAL ENZYME KINETICS

Many translated proteins are enzymes with specific metabolic functions in cells. Enzymes facilitate the reduction of the transition state energy required for a reaction to occur through various mechanisms. The processes involved are: maintaining the transition state, preventing water from entering the reaction site, stabilizing the substrate within the active site, and facilitating group transfer with the help of cofactors or coenzymes. The properties of the catalyst primarily determine the rate of enzyme-catalyzed reactions. The enzyme [E] catalyzes the initiation of a novel reaction pathway, similar to other catalysts. The unbound enzyme [ES] initially binds to the substrate [S].

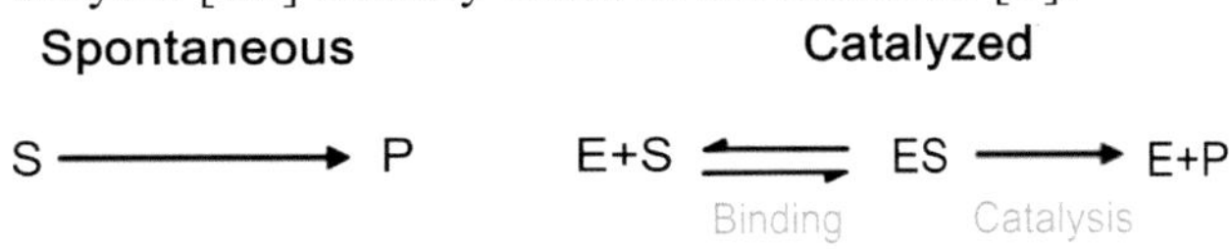

The Michaelis-Menten equation, which relates the initial velocity (vi) to the substrate concentration [S] and the parameters Km and Vmax, is employed to quantify the rate of this enzymatic process. The Vmax refers to the highest achievable velocity when the substrate concentration is infinite, while the Km represents the substrate concentration needed to reach half of the Vmax. The Michaelis constant (Km) is a measure of the enzyme's affinity for a substrate. Consequently, a low Km value arises when an enzyme has a strong attraction for a substrate, and the opposite is true as well (as shown in Figure 5.3).

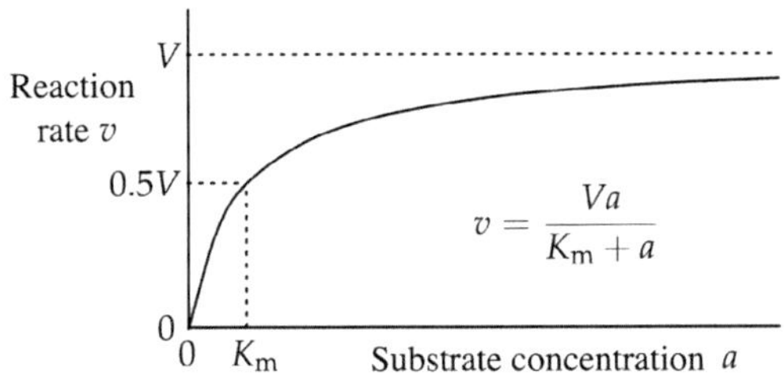

Figure 5.3. *Graphical depiction of the Michaelis–Menten equation.*

Source: Alvares (2024).

The Michaelis-Menten model makes several simplifying assumptions: substrate binding is in equilibrium, the production of [P] is irreversible, and [E] and [ES] are the only forms of enzymes.

Deducing reliable values for Vmax or Km from v versus [S] diagrams is difficult due to the asymptotic approach of vi towards Vmax (Figure 5.4). The Michaelis-Menten equation can be arranged in a manner such that the observed data points align in a linear fashion, facilitating the resolution of this issue (Ojha et al., 2019). The Lineweaver-Burk plot utilizes the technique of plotting the reciprocal of velocity (1/v) against the reciprocal of substrate concentration (1/[S]). The values for Vmax and –1/Km are determined by the points where the line of best fit intersects with the axes, as shown below.

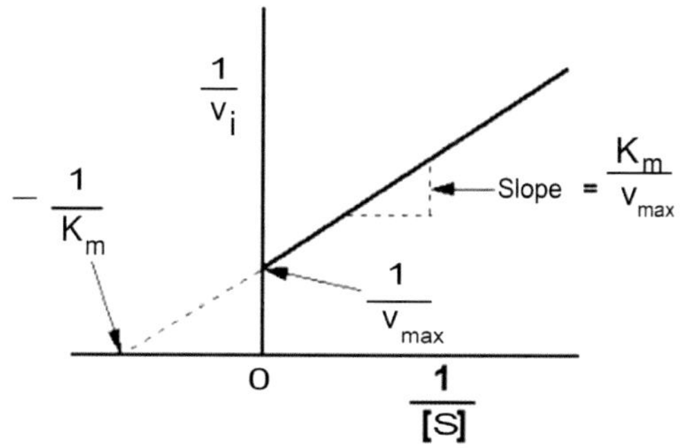

Figure 5.4. *A Lineweaver–Burk plot that serves to illustrate the values of Km and Vmax.*

Source: MetaFilter (2006).

5.2.1. Influences on the Rate of Enzyme Reactions

Several variables influence the catalytic properties of an enzyme, which consequently impact its activity. In order to establish a meaningful and reproducible measurement of an enzyme's activity, it is necessary to carefully adjust and control these parameters (Ojha et al., 2019). The variables encompass the concentrations of relevant substrates, cofactors, coenzymes, and inhibitors, along with physical parameters like temperature and pressure, as well as the chemical properties of the solution, such as pH and ionic strength. Cofactors, consisting of inorganic substances, and coenzymes, consisting of organic substances, often participate in reactions by either accepting or donating electrons or by temporarily stabilizing the substrate. The distinction between soluble coenzymes and prosthetic groups is determined by the type of interaction they have with the enzyme (Ojha et al., 2019). During a process, soluble coenzymes form a bond with the enzyme, undergo a chemical transformation, and subsequently detach from the enzyme. A subsequent, autonomous reaction restores the coenzyme to its initial state. Prosthetic groups, which are coenzymes, are firmly or covalently bound to an enzyme and remain associated with it throughout the entire reaction. After the initial release of the first product, the section of the substrate that is attached to the coenzyme is moved to another substrate or coenzyme of the same enzyme in a subsequent reaction. Animal cells lack the ability to synthesize numerous aromatic molecules, which results in the inability to produce coenzymes (Ojha et al., 2019). Metal ions can also serve as cofactors. Their functions are varied; some are responsible for preserving the natural conformation of an enzyme's active site, while others participate in oxidation-reduction reactions or polarize the chemical bonds in the substrate to facilitate catalysis.

5.2.2. Enzyme Regulation

5.2.2.1. Competitive and Noncompetitive Inhibition

When other substances interact with an enzyme, they can either inhibit or activate the enzyme. These factors will impact the reaction by modifying the reaction's Km or Vmax. Most enzyme inhibitors exhibit reversible effects and do not induce permanent alterations to the enzyme. In contrast, irreversible inhibitors have the ability to permanently and covalently modify the target enzyme. They are commonly known as suicide inhibitors (Ojha et al., 2019).

The comparison of the kinetics between the normal response and the inhibited response can be utilized to categorize inhibitors as either competitive or noncompetitive (Figure 5.5).

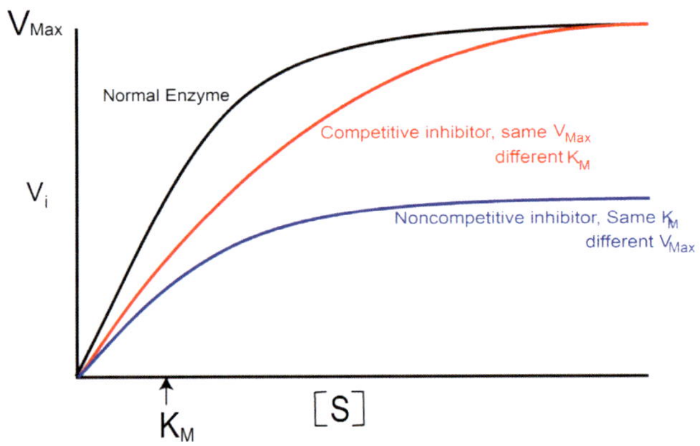

Figure 5.5. *Comparison between competitive and noncompetitive inhibition.*

Source: Maurya (2018).

Competitive inhibitors bind to the active site of an enzyme and compete with the substrate for binding. Many function as equivalents of substrates. The presence of an inhibitor causes an increase in the Michaelis constant Km, as a higher concentration of substrate is needed to achieve a rate that is half of the maximum. The maximum velocity (Vmax) will be achieved when the substrate concentration is raised to a level where the inhibitor is ultimately removed. Thus, competitive inhibitors do not influence the maximum rate, also known as Vmax (Ojha et al., 2019). In this case, increasing the substrate concentration can help overcome competition. However, it does not affect the maximum rate of reaction (Vmax). Nonetheless, the apparent Michaelis constant (Km) does increase because a higher substrate concentration is needed to reach Vmax. Contrarily, noncompetitive inhibitors attach to the enzyme at a distinct location from the substrate binding site, thereby preventing the enzyme from neutralizing their impact through the addition of more substrate. In this case, the maximum velocity (Vmax) decreases while the catalytic rate constant (kcat), which represents the rate of product production, remains constant (Ojha et al., 2019). The value of kilometers remains constant. Irreversible inhibitors generally cause noncompetitive inhibition due to the reduction in the concentration of active enzyme.

The Lineweaver-Burk plot effectively demonstrates the functioning of inhibitors. The x-axis intercept in this type of graph represents the value of −1/Km, while the y-axis intercept corresponds to 1/Vmax. This demonstrates that when a competitive inhibitor (A, red) is present and when there is no competitive inhibitor (blue), the straight lines intersect at the y-axis (1/ Vmax, remains the same). On the other hand, the straight lines obtained in the presence of noncompetitive inhibitors (B, red) have an unchanged x-intercept and a higher y-intercept (1/ Vmax increased, Km remains the same).

5.3. REGULATION OF ENZYMES IN ANIMALS

Animal live cells rely on the collaboration of numerous enzymes to function in a synchronized manner. Living cells do not synthesize or break down excess material that is not required for proper metabolism and growth (Gebert & MacRae, 2019). Transitioning between metabolic responses necessitates intricate regulatory mechanisms due to the complexity involved. Bacteria employ a wide range of methods to regulate the production and function of their enzymes. The regulation of gene expression encompasses the control of various steps involved in the transformation of a gene's activation into the ultimate production of an enzyme reaction.

Enzyme regulation is a vital biological process that controls the rate at which enzymatic reactions take place. This regulatory mechanism ensures the efficient functioning and adaptive response of metabolic pathways to the dynamic requirements of cells (Gebert & MacRae, 2019). Cells possess the ability to conserve energy, adapt to alterations in their environment, and uphold homeostasis by controlling enzyme activity. Enzyme activity is influenced by various factors such as temperature, pH, cofactors, regulatory molecules, and compartmentalization. Regulatory molecules have the ability to enhance (through activators) or diminish (through inhibitors) enzyme activity. These compounds specifically bind to enzymes, altering their catalytic activity. Moreover, the term "feedback inhibition" refers to the suppression of crucial metabolic enzymes by the product they control. To summarize, maintaining optimal cellular function and overall animal well-being necessitates accurate regulation of enzymes (Gebert & MacRae, 2019).

Regulation of animal enzymes is an intricate process that controls the functioning of enzymes to uphold optimal metabolic performance. Let us examine these systems:

- **Molecules that Regulate:** Enzyme-specific activator and inhibitor molecules possess the capacity to increase or decrease the activity of the enzyme.
- **Cofactors:** These are substances that are required for the proper functioning of enzymes. Enzymes require cofactors, which are non-protein molecules, in order to function properly.
- **Compartmentalization:** Enzymes can be safeguarded from damage or provided with optimal conditions for activation by being stored in specialized compartments.
- **Feedback Inhibition:** Feedback inhibition frequently inhibits key metabolic enzymes, resulting in the final output of the route.

5.3.1. Regulatory Molecules

Certain chemicals possess the capacity to regulate the functioning of enzymes, either enhancing or reducing their activity (Gebert & MacRae, 2019). Enzyme activators are compounds that enhance the activity of an enzyme, while enzyme inhibitors are substances that reduce the activity of an enzyme. Various chemicals can exert diverse effects on enzyme function, either by inhibiting or enhancing it.

- **Competitive vs. Noncompetitive:**Reversible binding refers to the temporary attachment of an activator or inhibitor molecule to an enzyme, which has been extensively documented in various well-studied cases. Specific classes of medication have the ability to act as reversible inhibitors. It inhibits the functioning of a viral enzyme, thereby impeding the virus's capacity to reproduce.

The classification of reversible inhibitors is based on their binding mechanisms (Gebert & MacRae, 2019).

An inhibitor can bind to the active site of an enzyme, thereby preventing the substrate from binding. Competitive inhibition occurs when an inhibitor competes with the substrate of an enzyme. Put simply, either the inhibitor or the substrate can be bound at a given moment.

In noncompetitive inhibition, the inhibitor does not hinder the substrate from binding to the active site. Instead, it attaches to a distinct site, thereby inhibiting the enzyme's ability to perform its function. The term "noncompetitive" is used to describe this type of inhibition where the inhibitor and substrate can bind simultaneously.

Gebert & MacRae (2019) suggest that the impact of competitive and non-competitive inhibitors on an enzyme's activity at different substrate concentrations can serve as a means to differentiate between them. When the amount of substrate is low, a competitive inhibitor will decrease the rate of reaction; however, it can be surpassed by a significant amount of substrate. Put simply, when there is an ample amount of substrate, the enzyme is capable of reaching its highest possible rate of reaction. During this occurrence, the substrate, rather than the inhibitor, will predominantly occupy the active sites in almost every molecule of the enzyme (Gebert & MacRae, 2019).

Despite a high concentration of substrate, an uncompetitive inhibitor will impede the enzyme-catalyzed reaction from reaching its maximum rate. The presence of a noncompetitive inhibitor renders the enzyme molecules incapable of carrying out their function, regardless of the amount of substrate available.

5.3.2. Allostery Regulation

Allostery is a term used to describe any form of regulation in which an activator or inhibitor binds to an enzyme at a location other than the active site. The allosteric site is the specific location where the regulator molecule binds (Gebert & MacRae, 2019). Figure 5.6 shows the allosteric inhibition and activation.

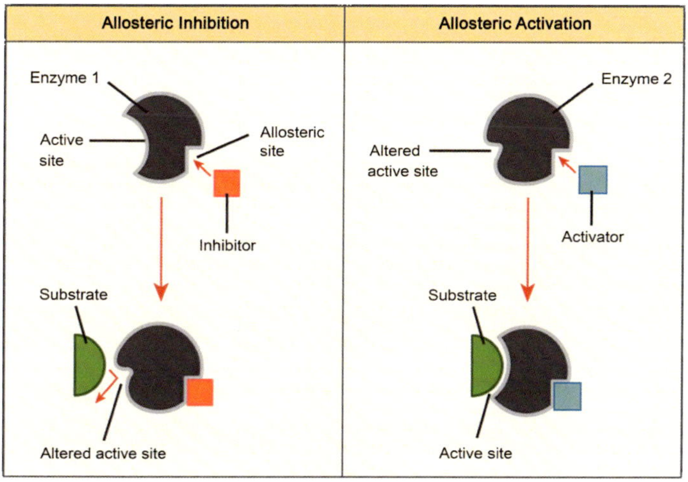

Figure 5.6. *The concept of allosteric inhibition and activation.*

Source: Enzyme Regulation (Article) | Khan Academy (2024).

Allosterically regulated enzymes can be identified by a distinct set of traits. These enzymes are commonly known as allosteric enzymes, and they include some of our most crucial metabolic regulators.

Allosteric enzymes typically possess multiple active sites that are distributed among different protein subunits. When an allosteric inhibitor binds to an enzyme, it causes minor modifications to the active sites on the protein subunits, resulting in decreased efficiency of the enzyme's function.

Allosteric activators represent an additional category. Allosteric activators can enhance the functionality of an enzyme's active site by binding to sites outside of the active site. In addition, the substrate can function as an allosteric activator through a phenomenon called cooperativity. Allosteric regulation, as defined by Gebert & MacRae (2019), occurs when the substrate affects active sites that are distant from its binding site.

5.3.3. Cofactors and Coenzymes

Enzymes that lack cofactors, which are non-protein molecules that assist them, exhibit reduced or no functional activity. These molecules have the ability to establish more robust covalent interactions that permanently bind them, or they can establish ionic or hydrogen bonds that temporarily bind them to the enzyme. Iron and magnesium are frequently found as cofactors in biological systems. Magnesium ions are essential for the functioning of DNA polymerase, the enzyme responsible for constructing DNA molecules (Gebert & MacRae, 2019).

Coenzymes are a type of cofactor that consist of organic compounds, meaning they are composed of carbon. Coenzymes are primarily derived from dietary vitamins. Certain vitamins act as coenzymes directly, while others serve as precursors to coenzymes. Vitamin C acts as a coenzyme for several enzymes that are involved in the production of collagen, an essential element of connective tissue (Gebert & MacRae, 2019). Figure 5.7 shows the structure of Vitamin C.

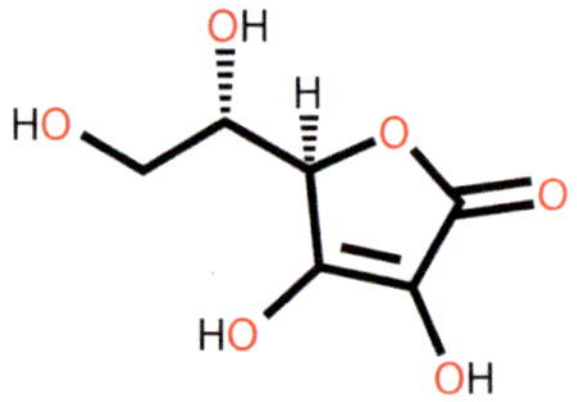

Figure 5.7. *The structure of vitamin C.*

Source: Devolli et al. (2021).

Enzyme compartmentalization refers to the process of organizing enzymes into specific compartments within a cell or organism. Enzymes are commonly segregated, meaning they are confined to a specific region within the cell, such as an organelle, where they carry out their functions. Compartmentalization enables the storage of enzymes in locations where they can efficiently carry out their specific activities. This arrangement facilitates easy identification of substrates, prevents damage to cells, and ensures optimal functioning conditions for the enzymes (Gebert & MacRae, 2019).

For instance, the lysosome's digestive enzymes work most effectively at a pH of approximately 5.0, which is found in the acidic interior of the lysosome but not in the cytoplasm, which has a pH of approximately 7.2. The low enzymatic activity of lysosomes at the cytosolic pH serves as a protective mechanism for the cell. In the event of a lysosome rupturing and releasing its contents, the enzymes will not initiate the breakdown of the cell due to the unfavorable pH conditions (Gebert & MacRae, 2019).

5.3.4. Feedback Inhibition of Metabolic Pathways

Feedback inhibition (Figure 5.8) refers to the process where the end product of a metabolic pathway inhibits the main enzyme that controls access to that pathway, thereby preventing the production of more of the end product.

It may initially seem peculiar why a molecule would intentionally obstruct its own pathway. The cell's ability to produce the precise amount of necessary substances is a remarkably intelligent strategy. If there is a shortage of product, the pathway will operate continuously to replenish the supply as the enzyme will not be obstructed. An excessive quantity of the product will obstruct the enzyme's function by accumulating, thereby impeding the production of new product until the available resources are depleted (Gebert & MacRae, 2019).

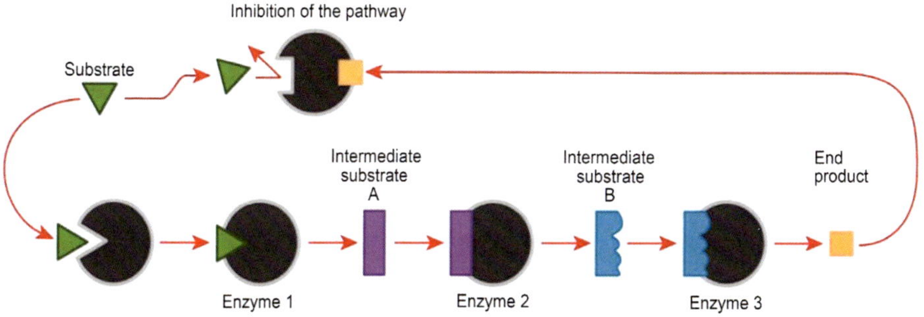

Figure 5.8. *Pathway of feedback inhibition.*

Source: Bartee et al. (2017).

Feedback inhibition typically occurs at the initial committed step of a pathway, which is the first step that is fundamentally irreversible. However, feedback inhibition may occur at various points along a pathway, particularly if it contains numerous branch points. Allosteric enzymes are commonly employed as catalysts for the process stages that are regulated by feedback inhibition (Gebert & MacRae, 2019).

Some enzymes involved in cellular respiration, which is the process that produces ATP to fuel cellular reactions, act as allosteric inhibitors of ATP, the energy carrier molecule. Feedback inhibition serves to prevent excessive ATP production in the presence of an excess of ATP. According to Gebert & MacRae (2019), the fact that ATP is an unstable chemical is beneficial. If an excessive amount of ATP is generated, a significant portion of it may be squandered and undergo spontaneous decomposition, reverting back to ADP and Pi.

In contrast, ADP functions as a positive allosteric regulator or activator for some of the enzymes that are inhibited by ATP. In order to enhance the activity of an enzyme, it is possible for ADP to bind to it and modify its structure (Gebert & MacRae, 2019).

5.4. ANIMAL ENZYME INHIBITION

An enzyme inhibitor is a chemical compound that binds to an enzyme and hinders its normal function. Enzymes, which are proteins, accelerate vital chemical reactions and convert substrate molecules into products. An enzyme facilitates a specific chemical reaction by binding the substrate to its active site, which is a specialized region that speeds up the most difficult step of the process.

An enzyme inhibitor disrupts this process through one of two mechanisms: either by binding to the active site of the enzyme and obstructing the substrate from binding, or by attaching to another location on the enzyme and impeding its ability to catalyze the reaction. Enzyme inhibitors can be bound to enzymes either permanently or reversibly (Amine et al., 2006). When irreversible inhibitors and an enzyme undergo a chemical bonding, the enzyme's activity is hindered until the bond is disrupted. Conversely, reversible inhibitors bind to enzymes without forming a chemical bond and can easily detach from the enzyme, allowing it to regain its normal functioning. Reversible inhibitors can cause different types of inhibition depending on whether they attach to the enzyme, the enzyme-substrate complex, or both (Amine et al., 2006).

Enzyme inhibitors are crucial for all cells because they are usually specific to a particular enzyme and help regulate its activity. For example, compounds produced at a later stage in a metabolic pathway can inhibit the activity of enzymes involved in that process, thereby reducing the production of unnecessary molecules. This type of adverse feedback is essential for maintaining cellular homeostasis. Furthermore, enzyme inhibitors play a crucial role in regulating proteases and nucleases, which are essential enzymes that can cause harm to a cell if not properly controlled. Enzyme inhibitors, substances that hinder the activity of crucial enzymes in prey or predators, are present in various toxins produced by plants or animals (Amine et al., 2006).

5.4.1. Mechanisms of Inhibition

** Diagrametic presentation is essential for each type of inhibition

5.4.1.1. Reversible Inhibitors

Reversible inhibitors form interactions with enzymes through ionic, hydrophobic, and hydrogen bonding mechanisms, which are all non-covalent in nature. The inhibitor forms multiple weak interactions with the active site of the enzyme, resulting in the formation of a strong and specific binding (Amine et al., 2006).

Unlike irreversible inhibitors, reversible inhibitors typically do not undergo chemical reactions when bound to the enzyme and can be easily removed through dialysis or dilution. Covalent reversible inhibitors are unique in that they have the ability to bind to the enzyme and break the connection, effectively reversing the inhibition (Amine et al., 2006).

5.4.1.2. Competitive

Simultaneous binding of both the substrate and the inhibitor is not possible in competitive inhibition. This phenomenon occurs when an inhibitor has a higher affinity for an enzyme's active site, where the substrate also binds, resulting in a competitive interaction between the inhibitor and the substrate for access to the active site. When the concentration of substrate reaches a sufficiently high level, the inhibitor can be surpassed in competition (while Vmax remains constant), resulting in the overcoming of this type of inhibition. However, due to the need for a higher concentration of substrate to reach the Km point, which is equivalent to half the Vmax, the apparent Km value will increase. Competitive inhibitors often have a similar structure to the actual substrate (Amine et al., 2006).

5.4.1.3. Non-Competitive

Non-competitive inhibition is a phenomenon where an inhibitor attaches to an enzyme, causing a decrease in its activity while not affecting the binding of the substrate. This type of inhibitor exhibits equal affinity for both the unbound enzyme and the enzyme-substrate complex. It can be regarded as having the ability to function as both uncompetitive and competitive inhibitors, without showing a preference for either. Therefore, the concentration of the inhibitor is the sole determinant of the extent of inhibition. The reaction will experience reduced efficiency, resulting in a decrease in Vmax. Nevertheless, the value of Km remains constant since the substrate will continue to bind appropriately.

5.4.1.4. Mixed Inhibition

This occurs when the inhibitor binds to the enzyme, depending on whether the substrate has already bound or not. Mixed inhibition refers to the combination of noncompetitive and competitive inhibition. Furthermore, variations in the inhibitor's binding strength to both the unbound enzyme and the enzyme-substrate complex may exist. Elevating substrate concentrations can partially alleviate this type of inhibition due to the competitive contribution, but it cannot entirely eliminate it due to the noncompetitive component. Although mixed-type inhibitors can potentially bind in the active site, the majority of cases of this type of inhibition are due to the inhibitor's allosteric effect, which occurs when it binds to a different location on the enzyme. When an inhibitor attaches to this allosteric site, it causes a change in the enzyme's conformation, specifically its three-dimensional shape or tertiary structure (Amine et al., 2006).

The effect of increasing the concentration of the substrate(s) on the extent of inhibition caused by a particular concentration of inhibitor can also be used to differentiate between these four types of inhibition. The level of inhibition intensifies as the concentration of substrate(s) rises in the case of uncompetitive (also referred to as anticompetitive) inhibition. It remains consistent for noncompetitive inhibition, and diminishes as the concentration of substrate(s) increases in the case of competitive inhibition (Amine et al., 2006).

5.5. ANIMAL-BASED ENZYME ASSAYS

Enzyme assays are employed to measure the quantity of a particular enzyme, which possesses well-defined characteristics, that is found in a liquid, partially purified sample, or tissue homogenate. In order for the assay results to have significance, it is crucial that the specimens are handled with utmost caution prior to the assay. Life is defined by the maintenance of cellular integrity and organization. Upon the death of an organism or the isolation of a tissue, autolysis, which are natural processes, initiate the destruction of cells, resulting in the breakdown of enzymes and cofactors. Autolysis is diminished when tissues are maintained at low temperatures. Therefore, specimens utilized in enzyme assays are usually stored at or below 4°C both prior to and following homogenate preparation (Haida & Hakiman, 2019).

Prior to conducting an enzyme assay, it is essential to store all specimens or preparations at low temperatures, as numerous enzymes are susceptible to rapid denaturation at elevated or even moderate temperatures. (Please note that these considerations are relevant for both the analysis of enzyme activity and the preparation of samples for studying enzyme characteristics).

It is evident that the assays should be conducted promptly, as the effects of autolysis and denaturation are likely to increase with time during storage. If a delay cannot be prevented, it may be crucial to store the specimens at extremely low temperatures, specifically –60°C, to maintain the enzyme activity. Nevertheless, there is no universally applicable protocol for storage as freezing tissues can potentially result in increased cellular disruption. It is imperative to evaluate the enzyme's stability under various temperatures, along with the properties of the material or preparation (Haida & Hakiman, 2019).

5.5.1. ENZYME ASSAY BY KINETIC DETERMINATION OF CATALYTIC ACTIVITY

To estimate the concentration of an enzyme in a preparation or fluid, the most effective approach is to measure its catalytic activity. This involves determining the amount of substrate that the enzyme can convert into product within a specific time frame and under predetermined conditions (Haida & Hakiman, 2019). As previously explained, the initial velocity of an enzyme-catalyzed reaction can be calculated using the equation v0 = k2[E0][S0]/([S0] + Km), where v0 represents the initial velocity, k2 is a constant, [E0] is the initial concentration of the enzyme, [S0] is the initial concentration of the substrate, and Km is the Michaelis constant. This equation is applicable when the Briggs-Haldane steady-state assumptions are valid. Here, k2 represents the rate constant for the production of the product, also known as kcat.

If the value of [S0] is fixed, the only variables in an assay are v0 and [E0]. Thus, the initial velocity (v0) is equal to a constant multiplied by the electric field strength (E0). In simpler terms, the speed at which an enzyme-catalyzed reaction starts is directly related to the amount of enzyme present. This relationship is linear when certain conditions are met, such as steady-state assumptions and a fixed initial substrate concentration (Haida & Hakiman, 2019).

Indeed, this relationship holds true with greater consistency when the substrate concentration is sufficiently high to approach saturation, although it remains theoretically valid even when the substrate concentration is fixed. There are two primary factors contributing to this. When all other factors are held constant, the integration of the appropriate rate equation shows that the duration of the linear, steady-state phase of the reaction increases as the degree of enzyme saturation increases. The reason for this is that as the value of [S0] increases, the rate at which the substrate is used becomes less important compared to the total concentration of substrate available (Haida & Hakiman, 2019).

As a result, higher values of S0 produce more precise estimates of v0 compared to lower values. Furthermore, the value of v0 is approximately equal to k2 multiplied by [E0], and v0 remains stable even when there are small variations in [S0] at high concentrations of [S0], as indicated by the equation ([S0] + Km) $\approx$ [S0]. This implies that if the concentration of [S0] is sufficiently high to nearly saturate the enzyme, consistent outcomes can be obtained from an assay system without the requirement of precisely fixing the substrate concentration. Thus, increasing the chosen value of [S0] enhances the feasibility of constructing a reliable assay system (Haida & Hakiman, 2019).

Referring to near-saturation is necessary instead of total saturation because an enzyme can only be completely saturated by substrate at an infinitely high substrate concentration, as mentioned in the Michaelis-Menten equation (Haida & Hakiman, 2019). Given the known Km value, this equation can be utilized to compute the v0/Vmax ratio for every initial substrate concentration. For example, if [S0] is equal to 0.5 Km and v0 is equal to 0.33 Vmax, or if [S0] is equal to 5 Km and v0 is equal to 0.83 Vmax, or if [S0] is equal to 50 Km and v0 is equal to 0.98 Vmax, these values would correspond. The v0/Vmax ratio can be used to determine the fractional saturation of the enzyme in a simple system (Haida & Hakiman, 2019).

As previously stated, it is clear that the ratios of v0 to Vmax and the values of fractional saturation are not affected by the concentration of [E0], as long as [S0] is significantly greater than [E0]. However, it is important to note that when measuring enzymes using kinetic methods, the values of v0 and Vmax vary depending on the initial enzyme concentration [E0]. The aforementioned explanation may create the perception that the process of enzyme assay can be perpetually enhanced by continuously increasing the substrate concentration. However, it is essential to take into account other factors as well. These factors include the solubility of the substrate, its price, and its potential for substrate inhibition. Hence, it is necessary to reach a compromise regarding the specific substrate concentration employed. In processes involving multiple substrates, it is necessary to maintain their concentrations at a fixed level, preferably close to saturation (Haida & Hakiman, 2019).

The concentration of a cofactor (or cofactors) can also impact the rate at which an enzyme-catalyzed process occurs. Just like substrates, cofactors also need to be present at a specific concentration, preferably in excess, to ensure a reliable and replicable enzyme assay system (Haida & Hakiman, 2019). An essential criterion for an effective assay system is the capacity to accurately monitor the catalyzed reaction, specifically by detecting alterations in optical, electrical, or other characteristics as the substrate undergoes transformation into the product. If this is not the situation, it may be possible to indirectly trace the path of the reaction by linking it to another reaction in which a similar change does take place (Haida & Hakiman, 2019).

5.6. ANIMAL ENZYME DYSFUNCTION DISORDERS

Enzyme deficiencies can also result in the development of disorders like maple syrup urine disease (MSUD) and phenylketonuria (PKU). Maple syrup urine disease (MSUD) arises from a scarcity of the enzyme branched-chain ketoacid dehydrogenase (BCKD), resulting in the buildup of amino acids that have the potential to harm nerves within the body. An identifiable symptom of MSUD is the existence of a pleasant, maple syrup-like scent in the urine. Conversely, PKU is a condition that arises from a lack of the enzyme phenylalanine hydroxylase (PAH), leading to elevated levels of phenylalanine in the bloodstream. Untreated PKU in childhood can result in cognitive impairment (Heindel et al., 2017).

Lysosomal storage disorders (LSDs) are another set of disorders that are connected to the functioning of enzymes. These disorders arise when the body is incapable of recycling cellular waste due to a deficiency of particular enzymes. The severity of lysosomal storage disorders (LSDs) is contingent upon the specific type and quantity of accumulated cellular waste. The majority of lysosomal storage disorders (LSDs) exhibit a gradual deterioration over time and have the potential to impact multiple organs and systems within the body (Heindel et al., 2017).

An instance of a lysosomal storage disorder (LSD) is Gaucher disease, which arises from a deficiency in the enzyme glucocerebrosidase. This enzyme is accountable for the degradation of a lipid compound known as glucocerebroside. Individuals with Gaucher disease experience the buildup of glucocerebroside in the cells of different organs, resulting in organ dysfunction and various symptoms including anemia, bone pain, and enlargement of the liver and spleen.

Pompe disease is an additional illustration of a condition that arises from a scarcity of the enzyme acid alpha-glucosidase. This enzyme is accountable for the degradation of glycogen, which is a stored form of glucose. Pompe disease is characterized by the accumulation of glycogen in cells, specifically in the muscles and heart. This results in muscle weakness, respiratory issues, and abnormalities in the heart (Heindel et al., 2017).

Enzyme deficiencies can also impact red blood cells, as observed in the case of glucose-6-phosphate dehydrogenase (G6PD) deficiency. G6PD is an enzyme that participates in the process of glucose metabolism. In individuals with G6PD deficiency, there is an insufficiency of the enzyme, resulting in the buildup of free radicals that have the potential to cause harm.

5.7. ROLE OF ENZYMES IN ANIMAL HEALTH

Enzymes are essential for maintaining animal health as they assist in the body's metabolic processes. Although the gastrointestinal microbiota of domesticated animals produce certain enzymes, not all enzymes are produced in adequate amounts to completely break down specific components of their diet, such as non-starch polysaccharides (NSP), proteins, and phytates. This can result in diminished nutrient assimilation and general health complications in animals (Choct, 2006). In order to tackle this problem, enzyme supplementation is frequently employed in animal feed to improve the quality of animals and feed, resulting in higher profits.

Enzymes can be incorporated into animal feed to enhance meat production efficiency and reduce costs, while also promoting the long-term well-being of the herd. Moreover, the utilization of enzyme supplements in animal feed has demonstrated a decrease in veterinary expenses and mortality rates (Choct, 2006). Enzymes can be obtained from diverse sources, such as animals, microorganisms, and concentrated plants. Enzymes are widely utilized in industries such as rubber, dairy, biofuel, farming, and brewing to enhance productivity and minimize costs (Choct, 2006).

When it comes to animal feed, the primary enzymes utilized are beta glucanases and arabinoxylanases. These enzymes degrade the fibrous components of grains, facilitating their digestion and enhancing the assimilation of vital nutrients in the stomach (Choct, 2006).

Utilizing enzymes in animal feed offers numerous advantages, encompassing economic, ecological, and physiological benefits. Enzyme supplements improve livestock health by enhancing the efficiency of nutrient absorption in the gastrointestinal tract. This enhances gastrointestinal health and hinders the proliferation of pathogenic microorganisms, resulting in improved animal health and decreased veterinary expenses (Choct, 2006).

Enzymes also contribute to the process of transitioning young animals from a diet of milk to solid food. Supplementing young animals' feed with enzymes can prevent gastrointestinal issues and decrease mortality rates, as they may not have enough enzymes for optimal digestion. The presence of incomplete digestion of a diet that does not include milk frequently leads to these problems, and the use of enzyme supplementation can assist in overcoming this difficulty (Choct, 2006).

Enzyme supplements in poultry farming have demonstrated the ability to enhance health, facilitate weight gain, and boost egg production in hens. Chickens can enhance their productivity and profitability by incorporating enzymes into their diet, resulting in increased meat or egg production (Choct, 2006).

The incorporation of enzymes in animal feed can yield substantial environmental advantages. One of the primary benefits is that it facilitates animals' proper food digestion, thereby decreasing manure production. This is significant because the waste produced by animals that can efficiently digest a larger portion of their food contains reduced levels of nitrogen and phosphorus.

Chapter 6

Animal Energy Metabolism

Contents

Energy is an essential resource that enables all living organisms to sustain their biological processes. The first law of thermodynamics, known as the law of conservation of energy, asserts that energy within organisms can only be transformed from one form to another, but cannot be generated or eliminated (Speakman, 2005). Some of the energy that enters an organism or reaches its surface is in a form that is not immediately advantageous, therefore not all of it provides direct benefits. The energetic efficiency of an organism can be quantified by comparing its energy consumption to its actual energy availability. Metabolism refers to the biochemical processes through which energy is converted from one form to another in living organisms. Metabolic processes, also known as energy conversions, occur in various tissues throughout the animal's body. An animal that is in motion has a heightened metabolic rate, while an inactive animal has a lower rate of metabolism (Speakman, 2005).

Cellular growth occurs when the rate of cell multiplication surpasses the rate of cell destruction. Animals experience weight loss when their cells undergo accelerated breakdown. Growth is caused by the assimilation of the by-products of digestion and rumen fermentation in ruminants into new body tissue. For example, within the rumen, amino acids undergo conversion into bacterial protein. The bacteria are subsequently metabolized into amino acids to facilitate absorption by the ruminant's body. Catabolism involves the breakdown of fat and carbohydrates molecules into smaller molecules of carbon dioxide and water. Urine comprises urea, a nitrogen compound that is excreted during the breakdown of protein molecules into carbon dioxide, water, and other compounds. The animal can now utilize the energy that was formerly stored in large molecules as a consequence of this process (Speakman, 2005).

Calories are a unit of measurement used to quantify the amount of energy produced by metabolism. A gram-calorie is approximately the amount of heat required to raise the temperature of one gram of water from 14.5°C to 15°C. The kilocalorie, being 1,000 times larger than the gram-calorie, is a more useful unit in physiological work.

6.1. BASAL METABOLISM AND RELATED TERMINOLOGY

Basal metabolism refers to the minimal energy expenditure of an animal when it is at rest in a thermoneutral environment and in a post-absorptive state (Speakman, 2005). The measurement does not accurately represent the minimal rate of metabolism necessary for sustaining life. To reduce the production of heat associated with food fermentation or nutrient metabolism, it is necessary to be in a post-absorptive state (Speakman, 2005).

The basal metabolic rate is sustained by the energy required for essential bodily functions such as respiration, circulation, excretion, secretion, and muscle tone maintenance. According to Speakman (2005), 75% of the energy required for basal metabolism is used for maintaining muscle tone and body temperature, while the remaining 25% is used for circulation, excretion, secretion, and respiration.

1. **Applications of Basal Metabolic Rate Measurements:**The basal metabolic rate (BMR) is a useful benchmark for comparing an animal's condition, whether it is physically active, on a different diet, ill, affected by parasites, pregnant, nursing, or experiencing any event that requires the use of energy. Therefore, it is beneficial for diagnosing conditions and evaluating how activity and output impact the energy requirements of a specific entity (Speakman, 2005).
2. **Criteria for Measurement:**Animals experience physiological responses to heat or cold stress, digestive processes, changes in activity, and psychological stress. Therefore, in order to establish a baseline for energy metabolism, it is necessary to provide standard conditions of thermoneutrality, post-absorptive digestion, a lying posture, and a calm psychological state. Thermoneutral conditions are necessary because an animal's metabolic rate increases in response to heat or cold stress. The thermoneutral range of most large animals is typically considered to be between 20 and 25°C, although this range can vary based on physiological, behavioral, and environmental factors. While air temperature can provide some insight into the temperature within a metabolic chamber, there are several other factors that need to be considered in a field setting (Speakman, 2005).

According to Speakman (2005), a post-absorptive digestive state is necessary as a standard condition due to the varying rates at which different diets are digested and produce heat. Research indicates that sheep experience a post-absorptive state lasting between 34 and 48 hours, during which food may still remain in their digestive tract. The duration of enhanced heat production is contingent upon the level of feeding. Sheep that were fed at a high level exhibited increased heat production during a seven-day fasting period.

When there is a lack of voluntary muscle movement, there may be an increase in heat generation. Both a physiological and psychological response to environmental stimuli may occur. One example of this is the common cold. The term "summit metabolism" refers to the highest metabolic rate achieved at a normal body temperature without any active muscle contractions (Speakman, 2005).

The term "resting metabolism" describes the metabolic rate when the body is not in a post-absorptive state, while "fasting metabolism" refers to the resting metabolism when the body is slightly post-absorptive. The term "basal metabolism" implies a minimum level of metabolic activity. When discussing their work, comparative physiologists should prefer using the term "standard metabolism." When determining the fasting period for ruminants, it can be difficult to accurately determine the exact point at which they enter a post-absorptive state. Therefore, fasting heat production is a suitable choice for measuring their metabolic rate (Speakman, 2005).

6.2. THE RELATIONSHIP BETWEEN HEAT PRODUCTION AND SURFACE AREA

1. **Heat Increments Due to Diet:** Animals generate more heat when they are being fed compared to when they are in a state after absorbing nutrients. The term "heat increment" denotes the difference in heat production between the act of feeding and fasting. The rise in temperature is a result of two processes: the breakdown of food in the digestive system and the production of heat during the conversion of nutrients into body tissue. During colder months, the animal highly values this heat energy as it helps maintain a balance between heat production and heat loss. The heat increment refers to the quantity of heat that must be dissipated in order to prevent an increase in body temperature, while other metabolic processes are sufficient to maintain this balance (Speakman, 2005).

The metabolic processes of both the individual animal and its recently born offspring can be influenced by dietary factors. A study suggests that insufficient prenatal nutrition may be associated with a lamb's reduced peak metabolism. Due to the need for rapid heat production after birth to withstand low temperatures, the lamb may experience mortality as a consequence. Adult sheep have been observed to experience a decrease in body temperature of 4°C when the ambient temperature dropped from 35°C to 20°C due to malnourishment. This unequivocally demonstrates the impact that food can have on maintaining homeothermy (Speakman, 2005).

2. **Reproductive Condition:** The reproductive cycle consists of three distinct phases: gestation, breastfeeding, and breeding. Each of these phases has a discernible effect on the energy metabolism.

Male ruminants expend a considerable amount of energy during the breeding season, resulting in a reduction in their consumption of forage and a noticeable decline in weight. The increase in energy expenditure is more likely due to the animal's increased overt activity, rather than increases in the rate of tissue metabolism itself (Speakman, 2005).

During pregnancy, the female organism must expend additional energy to sustain the uterus, nourish the developing fetus, fulfill the heightened requirements of her respiratory, circulatory, and excretory systems, and regulate the impact of hormones on her own metabolic processes. It is commonly recognized that a pregnant woman's energy metabolism does not increase in proportion to her weight until the last trimester of pregnancy, when fetal growth speeds up. However, some studies have indicated a slight decrease in metabolism during the first three months of pregnancy in cows. By quantifying metabolism based on body weight, even slight fluctuations in metabolic rate during the initial two-thirds of pregnancy can be accounted for. During the fast-growing fetal stage, an increase in metabolism may occur as the growth processes necessitate extra energy for the synthesis of fetal tissue (Speakman, 2005).

6.3. ATP SYNTHESIS IN ANIMALS

Adenosine 5′-triphosphate, also known as ATP, is an essential "energy molecule" found in all living organisms (Bonora et al., 2012). It is usually generated without the 5′-end. Specifically, it functions as a coenzyme that hydrolyzes the phosphate groups in enzymes such as ATP triphosphatase, thereby promoting the transfer of energy to cells. The molecule consists of three components: a triphosphate chain, a furanose ring, and an adenine bicyclic system.

In order to maintain their health and balance, animals need nourishment. Homeostasis refers to the ability of a system to maintain a stable internal environment in the face of external environmental fluctuations. The average human body temperature is typically 37°C (98.6°F). Regardless of whether the weather is extremely hot or freezing cold, humans are unable to alter the temperature. Food supplies the necessary energy to maintain the body at its current temperature. Carbohydrates, especially glucose, serve as the primary energy source for animals as they function as the body's fuel. By undergoing a series of catabolic chemical processes, digestible carbohydrates present in the food consumed by an animal are converted into glucose molecules and subsequently utilized as a source of energy (Bonora et al., 2012).

Cells utilize adenosine triphosphate, commonly known as ATP, as their primary energy source. Phosphate ester bonds serve as the storage site for energy in ATP. This energy is released upon the breaking of these bonds. ATP is converted into ADP and a phosphate group. Cellular respiration refers to the metabolic processes that proteins, lipids, and carbohydrates undergo in the cytoplasm and mitochondrion of the cell, resulting in the production of ATP (Figure 6.1).

Cells continuously utilize ATP hydrolysis for energy production, while also continuously synthesizing ATP from ADP and phosphate through cellular respiration processes. ATP synthase, an enzyme that converts ADP and phosphate into ATP, is primarily responsible for the production of ATP in cells. The cellular structures called mitochondria have a membrane that contains the enzyme ATP synthase (Bonora et al., 2012).

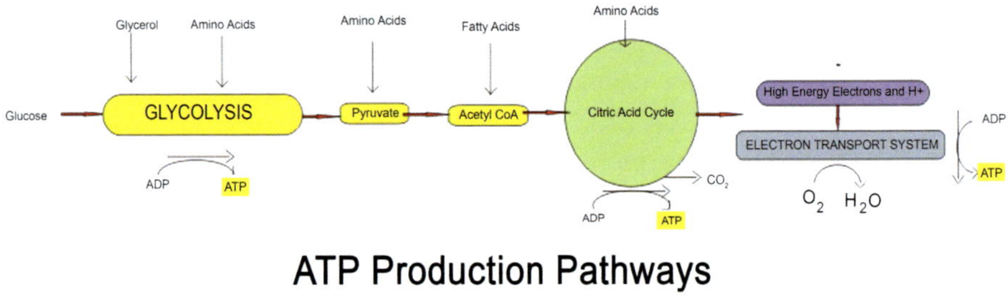

Figure 6.1.*The various pathways involved in the production of ATP.*

Source: Pressbooks (2020).

ATP is necessary for every biological process. It is utilized in the process of creating organic compounds necessary for the formation of tissues and cells. Furthermore, it provides energy for the transmission of electrical signals in the nervous system and the contraction of muscles. During times of increased ATP demand, the liver utilizes surplus ATP and glucose to synthesize glycogen, a polymerized form of glucose. This glycogen is then stored in both the liver and skeletal muscle cells. The liver synthesizes glucose from glycogen reserves in response to a decrease in blood glucose levels. Skeletal muscle undergoes glycogenolysis, converting glycogen into glucose, as a result of vigorous physical activity (Bonora et al., 2012). The conversion of glucose and excess ATP to glycogen is a crucial evolutionary advancement that facilitates the storage of additional energy.

- Animals obtain their energy from the consumption of food, which they subsequently utilize to maintain their body temperature and perform various metabolic processes.
- Animals consume glucose in their diets, which is converted into ATP, a form of energy, through cellular respiration.
- The liver converts excess ATP and glucose into glycogen, a substance that is stored for future use.

6.4. METABOLIC PATHWAYS INVOLVED IN THE PRODUCTION OF ENERGY

For animals to persist, a series of chemical processes known as metabolism occur within their cells (Luisa, 2012). Metabolism refers to the intricate network of metabolic pathways that work together to provide cells with the necessary energy for their functioning. It is apparent from Luisa's research in 2012 that these routes have remained unchanged over time, highlighting their significance to animals. In eukaryotes, the cytosol and mitochondria of cells are responsible for the metabolism of glucose or fatty acids, which is the main source of cellular energy in animals. Metabolism utilizes various metabolic pathways to either optimize energy absorption or minimize energy expenditure (Luisa, 2012).

Anabolism and catabolism are two distinct processes within metabolism that involve the synthesis and degradation of complex macromolecules, respectively (Figure 6.2). These processes consist of a series of chemical reactions. This text provides a comprehensive description of the fundamental metabolic processes necessary for sustaining life, as well as the molecular pathways that are responsible for generating and utilizing energy.

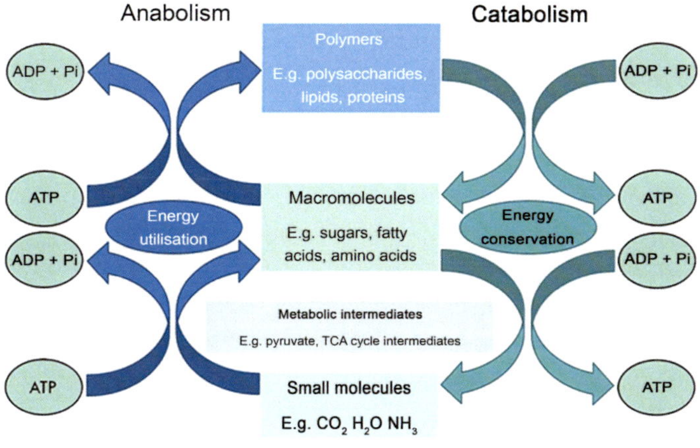

Figure 6.2. *The coupling of anabolic and catabolic pathways.*

Source: Judge et al. (2020).

Animals perpetually require energy to maintain cellular and overall bodily functions. Energy is required for the synthesis of intricate macromolecules such as glycogen, the transportation of ions and substrates like K^+, Mg^{2+}, and Ca^{2+} during heart contraction, and mechanical activities such as contraction and cellular movement.

6.4.1. Types of Energy

Hydrogen plays a crucial role in energy metabolism. Nicotinamide adenine dinucleotide (NAD^+) and flavin adenine dinucleotide (FAD) are hydrogen receptors that accept hydrogen from glucose during the process of glucose breakdown in animals ($C_6H_{12}O_6$). To release energy, the respiratory chain processes within the mitochondria oxidize hydrogen acceptors, which are also referred to as reducing equivalents. The synthesis of adenosine triphosphate (ATP) in living organisms is directly linked to the process of hydrogen oxidation (Luisa, 2012). ATP serves as the readily available energy source for cells, commonly referred to as the "molecular energy currency unit." ATP consists of three components: adenine, a nitrogenous base, ribose, a sugar, and triphosphate. The PO_4 bonds serve as energy storage units, and the breaking of each phosphate bond releases eight kilocalories of energy (Luisa, 2012).

6.4.2. Cellular Energy Production from Glucose: Metabolic Pathways

Cells utilize various enzymatic activities at different stages to convert ingested glucose into carbon dioxide and water. The catabolism of glucose involves two metabolic pathways: glycolysis and the tricarboxylic acid (TCA), also known as citric acid or Kreb's cycle (Luisa, 2012).

Glycolysis is a metabolic pathway that breaks down glucose into pyruvate. Glycolysis occurs in the cytosol of the cell, where the requisite enzymes for the process are present. Glycolysis is a metabolic process that converts glucose (a 6-carbon molecule) into two 3-carbon end products called pyruvates. In anaerobic metabolism, these pyruvates are further converted into lactic acid. The process is an oxidative catabolic process that generates reduced NAD (NADH) and ATP. Fructose and galactose undergo catabolism through the process of glycolysis, wherein they are converted into glycolysis intermediates. Hexokinase possesses the capacity to transform fructose into fructose-6-phosphate. Galactose can be incorporated into glycolysis by undergoing a series of conversions. First, it is converted to galactose-1-phosphate. Then, galactose-1-phosphate is transformed into glucose-1-phosphate, which is further converted to glucose-6-phosphate (G6P). G6P serves as an intermediate molecule in the process of glycolysis (Luisa, 2012).

6.4.3. Glycolysis

Glycolysis is a biphasic process that encompasses the generation of ATP in the energy-yielding phase and an energy-requiring phase that necessitates the utilization of ATP (Figure 6.3). Hexokinases, an enzyme present in cells that consume glucose, catalyze the phosphorylation of glucose, resulting in the attachment of a phosphorus group and the conversion of glucose into G6P through the utilization of ATP. The cellular "machinery" can now initiate the process of metabolizing glucose. Hexokinase facilitates the transfer of phosphate from ATP to glucose during the initial stage of glycolysis, leading to the creation of glucose-6-phosphate. Therefore, ATP is utilized in this stage to provide the necessary energy for the reaction to persist (Luisa, 2012).

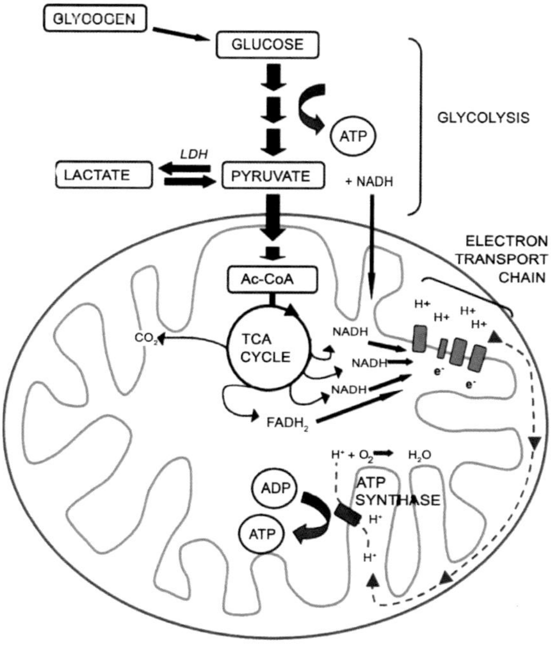

Figure 6.3.*The process of glycolysis occurring in the cytosol.*

Source: Sarvasmaa (2007).

Luisa (2012) states that the conversion of glucose-6-phosphate to fructose-6-phosphate and then to fructose-1,6-biphosphate results in the production of dihydroxy acetone phosphate (DHAP) and glyceraldehyde-3-phosphate (G3P) by breaking down the latter. An additional ATP is required for this process in order to add a phosphate group to the intermediate fructose-6-phosphate. Consequently, during the "preparation" phase, the metabolism of each glucose molecule necessitates the use of two ATP molecules (Luisa, 2012). Pyruvate is formed through the subsequent processing of G3P during the payoff phase. During the intermediate steps of this phase, one molecule of NADH and two molecules of ATP are produced. The DHAP produced can be readily converted into G3P and is managed in a manner similar to the original G3P. According to

Luisa (2012), each glucose molecule will yield two NADH molecules, four ATP molecules, and two pyruvate molecules.

The terms TCA, Citric Acid, and Kreb's Cycle refer to the same metabolic pathway in cells. The TCA cycle, also referred to as the citric acid cycle or Kreb's cycle (named after Hans Krebs, the discoverer in 1937), is a series of chemical reactions occurring in the mitochondrial matrix and facilitated by enzymes (Figure 6.4). Several enzymes that contain B vitamins, such as thiamine, riboflavin, and niacin, function as coenzymes during the process. The TCA cycle is a crucial element of aerobic respiration in cells (Luisa, 2012). The TCA cycle in the animal body produces precursors for the storage of lipids (fuels) and amino acids (building blocks).

As stated earlier, pyruvic acid undergoes pyruvate dehydrogenase (PDH) conversion to acetyl CoA prior to entering the cycle. To participate in the cycle, acetyl CoA (consisting of 2 carbon atoms) must react with oxaloacetate (consisting of 4 carbon atoms) to form citric acid (consisting of 6 carbon atoms). This response propels the cycle forward. Approximately 10 distinct enzymatic reactions convert citric acid into various intermediate compounds, such as fumarate, malate, succinate, and α-ketoglutarate.

During various processes, NAD receives high-energy electrons, and NAD molecules also acquire H^+ to generate NADH (Luisa, 2012). FAD serves as the recipient of electrons in a particular phase, assimilating H^+ ions to generate $FADH_2$. One molecule of acetyl CoA is also utilized to generate one molecule of ATP in one of the metabolic pathways. The ultimate outcome of the cycle is the production of oxaloacetic acid, which is identical to the acid present at the beginning of the cycle.

This acid will then combine with another acetyl CoA molecule to initiate the cycle once more. The TCA cycle produces a total of six NADH, two $FADH_2$, two ATP molecules, and either one mole of glucose or two moles of pyruvic acid. The electron transport chain utilizes both FADH and NADH to generate ATP. According to Luisa (2012), the TCA cycle produces a total of 24 ATP per mole of glucose.

When a glucose molecule is fully metabolized into CO_2 and H_2O, we obtain a total of 38 ATP. Each ATP high-energy phosphate bond possesses a total of eight kilocalories of energy. Hence, the net energy recovered amounts to 304 kcal, calculated by multiplying 38 by 8. The efficiency of converting glucose bond energy into ATP high-energy P bond is 45% calculated as 304/674*[1] × 100.

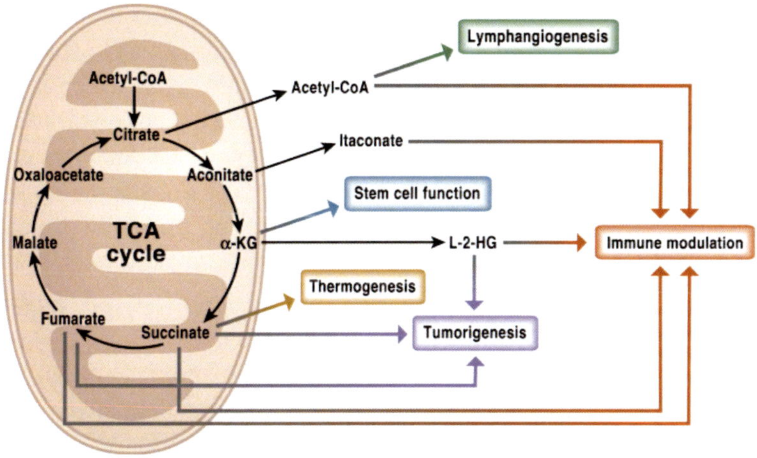

Figure 6.4. *Tricarboxylic acid (TCA) cycle occurring in the mitochondria.*

Source: Martínez-Reyes & Chandel (2020).

The conversion of oxaloacetate and acetyl CoA into citrate is the rate-limiting step in the TCA cycle. In order to maintain cell viability and produce ATP, it is necessary to have adequate levels of oxaloacetate. Nutritional disorders such as ketosis, which result from inadequate energy production, require the utilization of oxaloacetate as a precursor to initiate the tricarboxylic acid (TCA) cycle. Furthermore, the TCA cycle produces intermediates that serve as precursors for the biosynthesis of amino and fatty acids (Luisa, 2012). Oxaloacetate can be used for the production of nonessential amino acids, specifically glucogenic amino acids and most nonessential amino acids like alanine, serine, cysteine, and glycine. Citrate, on the other hand, is utilized for the synthesis of fatty acids. Amino acids have the ability to produce glucose when glucose levels are low, and they can also be utilized to synthesize glutamic acid through α-ketoglutarate (Luisa, 2012).

6.5. REGULATION OF ENERGY METABOLISM

The regulation of energy metabolism is complex to maintain a balance between energy production and consumption, ensuring efficient utilization of nutrition and adaptability to changing physiological needs (Nakamura et al., 2014).

6.5.1. Feedback Mechanisms

Feedback systems regulate enzyme activity to maintain homeostasis in energy metabolism by detecting changes in metabolic intermediates or end products (Nakamura et al., 2014). One of the classic examples is the regulation of glycolysis by the availability of ATP and citrate. When energy levels reach a certain threshold, large quantities of ATP and citrate inhibit phosphofructokinase, a crucial enzyme in glycolysis, thereby reducing the rate at which glucose is broken down. Conversely, phosphofructokinase activity is enhanced by low ATP levels and high ADP levels, promoting glycolytic flux for increased ATP production. Within other metabolic pathways, similar feedback systems ensure the appropriate utilization of nutrients and prevent metabolic imbalances (Nakamura et al., 2014).

Hormonal regulation is crucial for coordinating energy metabolism in response to physiological cues and environmental signals. Insulin and glucagon are two crucial hormones that regulate glucose metabolism. When blood glucose levels rise, the pancreas secretes insulin, which promotes the absorption of glucose by cells, the formation of glycogen in the muscles and liver, and the generation of lipogenesis in adipose tissue. Conversely, to regulate blood glucose levels, glucagon is released during periods of fasting or low glucose, stimulating gluconeogenesis and glycogen breakdown in the liver (also called glycogenolysis). In addition, hormones such as growth hormone, cortisol, and epinephrine (adrenaline) play a role in mobilizing energy stores, enhancing substrate availability, and regulating metabolic rate in response to stress, exercise, or fasting. These hormones also impact energy metabolism (Nakamura et al., 2014).

6.5.2. Cellular Signaling Pathways

Cellular signaling pathways play a crucial role in regulating energy metabolism at the cellular level by integrating various signals from hormones, nutrients, and environmental stimuli. The AMP-activated protein kinase (AMPK) pathway plays a significant role in regulating energy metabolism. AMPK is activated as a cellular energy sensor in response to an increase in the AMP/ATP ratio, indicating a decline in cellular energy levels. AMPK promotes energy generation and conservation by phosphorylating specific proteins involved in mitochondrial biogenesis, fatty acid oxidation, and glucose uptake. Conversely, when there is sufficient energy and nutrition, the mammalian target of rapamycin (mTOR) signaling pathway, which is activated by growth hormones and nutrients, stimulates anabolic processes such as protein synthesis and cell growth (Nakamura et al., 2014).

6.5.3. Analysis of Nutrient Sensing and Metabolic Adaptation

Cells possess intricate mechanisms to detect the presence of nutrients and adjust metabolic processes accordingly (Nakamura et al., 2014). Various mechanisms in the body, such as sirtuin family proteins (SIRTs) and the mechanistic target of rapamycin complex 1 (mTORC1), respond to changes in nutrient levels, such as amino acids, glucose, and fatty acids, in order to control cellular metabolism (Nakamura et al., 2014). These pathways regulate crucial metabolic processes such as autophagy, protein synthesis, and mitochondrial activity in response to cellular stress, energy levels, and nutrient availability.

Environmental factors such as temperature, oxygen availability, and circadian rhythms have an impact on energy metabolism. Cold exposure stimulates thermogenesis and lipid metabolism to generate heat and maintain body temperature, while low oxygen levels, known as hypoxia, induce adaptive responses that enhance oxygen transportation and metabolic efficiency. The hypothalamic master circadian clock regulates circadian rhythms and determines the timing of metabolic processes, such as glucose and lipid metabolism, in relation to the light-dark cycle (Nakamura et al., 2014).

6.5.4. Influences on Energy Metabolism

1. **Nutritional Factors:** The nutritional composition has a significant impact on energy metabolism. Macronutrients, such as proteins, lipids, and carbohydrates, serve as the primary constituents for the production of energy. The composition and quality of different macronutrients in the diet impact energy equilibrium, substrate utilization, and metabolic pathways. Carbohydrate-rich diets stimulate the production of insulin and glycolysis, which promote the uptake of glucose and its conversion into fat or glycogen for storage. Conversely, diets that are rich in fat and low in carbohydrates lead to a shift in metabolism, resulting in the production of energy through the breakdown of fatty acids and the creation of ketones. Protein consumption impacts energy expenditure by preserving lean body mass, which increases basal metabolic rate, and by influencing the thermic effect of food (Nakamura et al., 2014).
2. **Hormonal Regulation:** Hormones play a crucial role in regulating energy metabolism by affecting hunger, energy expenditure, and substrate utilization. The hormones insulin, glucagon, leptin, ghrelin, and adiponectin play crucial roles in regulating energy balance, lipid metabolism, and glucose homeostasis. To maintain stable blood glucose levels during fasting, the hormone glucagon stimulates the breakdown of glycogen and the production of new glucose, while insulin promotes the uptake and storage of glucose in muscles and adipose tissue. Leptin, which is generated by adipose tissue, communicates the feeling of being full and enhances thermogenesis, thereby regulating appetite and energy usage. Dysregulation of

hormonal signaling pathways can lead to metabolic diseases and imbalances (Nakamura et al., 2014).

3. **Environment Factors:** Several environmental factors have the potential to influence energy metabolism and metabolic health. These factors include stress levels, light exposure, sleep patterns, and ambient temperature. Cold exposure stimulates brown adipose tissue activity and thermogenesis, leading to increased heat production and regulation of body temperature. Conversely, chronic stress and sleep disorders disrupt circadian rhythms, hormone equilibrium, and appetite regulation, thereby worsening metabolic dysfunction. Environmental pollutants, endocrine disruptors, and dietary toxins not only interfere with metabolic pathways and disrupt hormone signaling but also increase the likelihood of developing metabolic diseases (Nakamura et al., 2014).

6.6. ANIMAL ENERGY METABOLISM DISORDERS

Storage disorders and inherited metabolic abnormalities can be classified into two main categories: acquired or genetic. These disorders are characterized by the accumulation or storage of specific substances or byproducts of lysosomal enzymes within cells, due to a partial or complete deficiency of certain enzymes. Lysosomal storage diseases primarily manifest symptoms in the central nervous system (CNS), although they often impact extensive regions of the body.

1. **Genetic Storage Diseases in Animals:** The nomenclature for genetic storage disorders is based on the specific metabolic waste product that accumulates in the lysosomes. Typically, animals exhibit no abnormalities at birth and begin displaying symptoms within the initial weeks or months of their existence. Due to the absence of targeted treatments for these diseases, they have a tendency to progress and ultimately lead to mortality (Heindel et al., 2017).

Siamese, Korat, and domestic cats, as well as Beagle crosses, German Shorthaired Pointers, Japanese Spaniels, Portuguese Water Dogs, English Springer Spaniels, Alaskan Huskies, and Shiba Inus, are prone to developing gangliosidoses (GM1 and GM2), which are a group of disorders affecting small animals. Siamese and domestic shorthair cats, along with German Shepherd and Poodle dogs, are susceptible to sphingomyelinosis. Australian Silky Terriers and Dalmatians are susceptible to developing glucosecerebrosidosis. Certain breeds of domestic cats and dogs, including English Setters, Cocker Spaniels, Dachshunds, Chihuahuas, Salukis, and Border Collies, have the potential to develop a condition called ceroid lipofuscinosis. Mannosidosis can be contracted by both Persian and domestic cats. Silky Terriers, domestic shorthaired cats, and Norwegian forest cats are prone to glycogenosis. Cairn Terriers, West Highland White Terriers, Beagles, Bluetick Hounds, Poodles, and domestic shorthair cats are susceptible to developing globoid cell leukodystrophy, commonly referred to as Krabbe disease. Siamese cats are prone to mucopolysaccharidosis type IV, whereas Korat and domestic shorthair cats are prone to type I (Heindel et al., 2017).

Plott Hounds, Rottweilers, Labrador Retrievers, Wirehaired Dachshunds, New Zealand Huntaways, Schipperkes, Welsh Corgis, Miniature Schnauzers, Chesapeake Bay Retrievers, German Shepherd Dogs, Rat Terriers, and mixed-breed dogs are susceptible to developing mucopolysaccharidosis, a condition associated with lameness (Heindel et al., 2017). Pyruvate kinase deficiency is a condition found in Basenjis, Beagles, West Highland White Terriers, and Cairn Terriers. Phosphofructokinase deficiency is observed in English Springer Spaniels and American Cocker Spaniels. Porphyria is a disease that affects Siamese and domestic shorthair cats. These conditions are associated with reduced red blood cell survival and the development of anemia.

Alpha-mannosidosis impacts the health of various breeds of large animals, such as Holstein, Angus, Simmental, Galloway, and Murray Grey cattle. Goats that have been bred with Nubians and Saler cattle are prone to beta-mannosidosis. GM1, also known as generalized glycogenosis, is a condition that specifically impacts Suffolk sheep and Holstein cattle. Pigs, Shorthorn cattle, and Brahman cattle are prone to generalized glycogenosis (GM2). Globoid cell leukodystrophy has been observed in polled Dorset sheep (Heindel et al., 2017).

Neuronal lipodystrophy, shaker calf syndrome, maple syrup urine disease, and hereditary neuraxial edema are inherited diseases found in Angus, Beefmaster, horned Hereford, polled Shorthorn, and Hereford-Friesian cross cattle (Heindel et al., 2017). No cases of lysosomal storage diseases have been documented in horses. However, there are inherited diseases that cause neurological symptoms, such as congenital encephalomyelopathy in Quarter Horses and inherited myoclonus in Peruvian Paso foals (Heindel et al., 2017).

Additional hereditary diseases that involve fundamental errors of metabolism in different tissues include goiter in sheep and goats, inherited parakeratosis (edema disease) in cattle, osteogenesis imperfecta in sheep and cattle, potentially cardiomyopathy in cattle, hypotrichoses, baldy calves, photosensitization in sheep, dermatosis vegetans in pigs, porcine stress syndrome, dermatochalasis and Ehlers-Danlos syndrome in cattle, hemochromatosis in cattle, and Marfan syndrome in cattle. Various hereditary disorders are also prone to involve errors in the metabolism of structural tissues, specifically those that rely on the improper development of collagen, cartilage, and bone. Several metabolic issues have been recorded in association with dysfunctions of the immune system (Heindel et al., 2017).

2. **Animal Storage Diseases:**Ingesting plants that contain inhibitors of lysosomal catabolic enzymes can lead to the development of acquired storage disorders. Consuming Locoweed plants (Astragalus or Oxytropis spp.) for a long time can lead to the development of acquired brain storage illness. Several toxic compounds, including indolizidine alkaloids, swainsonine n-oxide, and locoine, inhibit the function of alpha-mannosidase. Horses are the primary animals that experience intoxication, although cattle, sheep, and goats can also be affected (Heindel et al., 2017).
3. **Mitochondrial Disorders:**These hereditary disorders are characterized by dysfunctions in mitochondrial function, leading to reduced energy production and cellular malfunctions. Malfunctions in oxidative phosphorylation, mitochondrial biogenesis, and mitochondrial dynamics result in the disruption of energy metabolism in diseases related to the mitochondria. Mutations in genes that encode nuclear-encoded mitochondrial proteins, mitochondrial DNA (mtDNA), and mitochondrial proteins are responsible for mitochondrial dysfunction and metabolic abnormalities. Various clinical manifestations, including muscle weakness, exercise intolerance, neurological impairments, cardiomyopathy, and dysfunction of multiple organ systems, are characteristic of mitochondrial diseases (Heindel et al., 2017).

6.7. ENERGY EQUILIBRIUM IN ANIMALS

Obesity is a clear consequence of the inadequate regulation of energy balance in both humans and animals. Gaining insight into the mechanisms by which animals regulate their energy balance or metabolism will facilitate comprehension of the development of obesity and potentially enable its prevention. However, placing exclusive emphasis on obesity may lead to an excessive diversion of attention from studying the adipose tissue, which serves as an indicator of a predisposition towards positive energy balance but not necessarily the underlying cause of it (Hill et al., 2012).

An animal is said to be in energy balance when there is no change in its body's energy reserves. Subsequently, the amount of energy he expends (in the form of heat) matches the amount of energy he consumes (from metabolizable food). If the amount of energy consumed is greater than the amount of energy expended, the energy balance is positive. Conversely, if the amount of energy expended is greater than the amount of energy consumed, the energy balance is negative (Hill et al., 2012).

Controlling energy intake is considered the main determinant of energy balance regulation due to its high variability in the equation. This can be comprehensible considering the multitude of difficulties linked to accurately measuring energy expenditure over an extended period of time. Various regulatory mechanisms for energy intake have been proposed. Booth further elaborated on these concepts by creating computer models that could forecast the feeding patterns of both rats and humans. The researchers hypothesized that the transmission of signals to the neural receptors responsible for regulating feeding behavior is contingent upon the availability of energy sources for metabolic processes, such as sugars, amino acids, ketones, glycerol, and lactate (Hill et al., 2012).

1. **The Function of Adipose Tissue in Maintaining Energy Equilibrium:**Multiple pieces of data provide support for the notion that genetics plays a determining role in the mechanisms of fat metabolism in adipose tissue. Pigs bred for leanness demonstrate a more efficient mobilization of fatty acids in response to lipolytic hormones compared to pigs selected for fatness. There is evidence indicating that obesity in humans can be attributed to a genetic predisposition that leads to a disruption in the process of fat mobilization. The development of larger fat cells, which are characteristic of obese animals, is associated with a decreased responsiveness to lipolytic hormones. However, there is limited evidence to suggest that a primary deficiency in adipose tissue is a major contributing factor in the development of obesity (Hill et al., 2012).

The transplantation of tissue from genetically thin or obese "donor" mice into lean or obese "hosts" led to the fat cells adopting the characteristics of their new hosts. This indicates that the tissue environment plays a significant role in determining the size of fat cells. The total size of a fat depot is determined by the number and dimensions of its fat cells. At three weeks of age, rats that were fed in small litters, also known as "overnutrition," showed larger epididymal fat pads compared to rats that were fed in large litters, also known as "undernutrition." A study conducted by Hill et al., in 2012 discovered that the adipose tissue of severely obese adults, who had been obese since childhood, exhibited an augmentation in both the quantity and dimensions of adipocytes. These observations led to the formulation of two hypotheses: (1) hypercellularity is a consequence of early nutritional experience; and (2) hypercellularity is a factor that contributes to the perpetuation of obesity (Hill et al., 2012). Insufficient evidence exists to substantiate either aspect of the hypothesis. The insufficient methodologies employed to "quantify" adipocytes render it unfeasible to ascertain whether the quantity of adipocytes has increased. Undeveloped adipocytes that have not accumulated sufficient triacylglycerol are not identifiable, but they have the potential to accumulate fat in the future, leading to a misleading increase in the count of identified adipocytes (Hill et al., 2012).

6.8. THE SIGNIFICANCE OF ENERGY METABOLISM IN MAINTAINING ANIMAL HEALTH

Energy is essential for the maintenance of the physical structure, reproduction, and adaptation of all living organisms. Metabolism, a collection of chemical reactions, enables organisms to convert the chemical energy stored in molecules into a form of energy that can be used by their cells. Animals metabolize the carbohydrates, fats, proteins, and nucleic acids in their food to obtain chemical energy for their bodily functions. Plants convert solar light energy into chemical energy stored in molecules through the process of photosynthesis (Aminzadeh-Gohari et al., 2019).

6.8.1. Chemical Reactions and Bioenergetics

Bioenergetics refers to the scientific concept that explains the flow of energy within living organisms, such as cells. Comprehensive chemical reactions are the processes through which intricate compounds are constructed and disassembled in cellular activities. While certain chemical reactions necessitate energy for their progression, others occur spontaneously and release energy. The cellular metabolism encompasses all chemical processes that take place within a cell, including both energy-releasing and energy-consuming reactions (Aminzadeh-Gohari et al., 2019).

In animals, metabolism refers to the collective set of physical processes that involve the absorption, breakdown, transformation, excretion, and utilization of food. Metabolic complications and diseases can arise as a consequence of this phenomenon. Carbohydrates from food are metabolized by the body into glucose. Glucose serves as the main energy source for regular metabolic processes. Excess glucose is stored as glycogen in the muscles or liver and can be readily released. Additionally, it circulates throughout the body and is accessible to all cells. The crucial component for this procedure is the hormone insulin, which is produced by the pancreatic beta cells and secreted into the bloodstream. The liver regulates the release of glucose and facilitates the transportation of glucose into muscle, fat, and liver cells. Insulin regulates blood glucose levels in this way. No other hormone is capable of performing this function (Aminzadeh-Gohari et al., 2019).

The study conducted by Du et al. (2017) aimed to examine the influence of different levels of dietary energy on the quality of yak meat and the microorganisms found in the rumen. The researchers discovered that yaks that received a substantial amount of dietary energy had elevated levels of water and crude fat in their longissimus pectoris muscle, suggesting enhanced muscle quality. The enhancement in meat quality was mainly ascribed to the augmented prevalence of ruminal amylolytic bacteria, which contribute to the degradation of carbohydrates and furnish the essential components for the synthesis of fatty acids. In a separate investigation conducted by Kong et al. (2017), the scientists specifically examined alterations in the metabolic processes of dairy cows residing at varying high elevations. The researchers examined blood biochemical markers and employed metabonomics to identify the specific metabolites linked to altitude. The findings indicated that the metabolites that exhibited an elevation in concentration with increasing altitude were predominantly linked to sphingolipid and amino acid metabolism. An inverse correlation was observed between higher altitude and sphingolipid metabolism. The scientists discovered a distinct metabolic pathway in the mammary glands of dairy cows experiencing low oxygen levels, which was linked to the degradation of glycerophospholipids. The metabolic abnormality ultimately resulted in lipid abnormalities and the death of mammary epithelial cells in cows. Qian et al. (2017) conducted a study investigating the effects of vitamin E deficiency in cows during the early postpartum period. A deficiency in vitamin E resulted in a notable reduction in the concentrations of specific proteins in the bloodstream, such as serum amyloid protein A4, apolipoprotein A3, and pantetheinase-1. Pantetheinase-1 exhibited a specific correlation with subclinical vitamin E deficiency in dairy cows and holds promise as a biomarker. Mesgaran et al. (2017) examined the impact of rumen-protected l-carnitine on high-yielding dairy cows in the perinatal period. The researchers discovered that l-carnitine increased production, enhanced liver metabolism, and regulated health biomarkers in the cows. The authors proposed additional investigation to ascertain the influence of l-carnitine nutrition during pregnancy on the uterus and its implications for the health and performance of the offspring. Wang et al. (2017) investigated the influence of estrogen on the regulation of glucose metabolism and energy balance. It was discovered that estrogen increased the synthesis and storage of glycogen, controlled autophagy, and improved the absorption of glucose. Estrogen was discovered to suppress apoptosis and boost the immune function of polymorphonuclear (PMN) cells. Nevertheless, the precise mechanism by which estrogen controls glucose metabolism remains incompletely comprehended.

Chapter 7

Hormones and Signaling in Animal Biochemistry

Contents

Hormones and signals in animal biochemistry play a crucial role in coordinating essential functions in living organisms (Krauss, 2006). Signaling molecules, known as hormones, are present in both plants and animals. They exert their effects on organs and tissues located far from their site of production by traveling through the bloodstream. When a cell or gland secretes a hormone, it transmits precise signals to other body parts, effectively modifying cell metabolism. Hormones, even in small quantities, have a significant impact by acting as messengers between cells. The endocrine system, composed of various glands, releases these hormones into the bloodstream, where they exert specific effects on target tissues. Hormonal actions are diverse and crucial for maintaining homeostasis, encompassing functions such as immune system modulation and growth regulation (Krauss, 2006). This chapter will provide a comprehensive examination of hormones and signaling in animal biochemistry, including an overview of hormones in animals, the synthesis and regulation of hormones, hormone signaling pathways, the endocrine system in animals, disorders of hormone function in animals, the role of hormones in animal health, and future directions in hormone research.

7.1. A BRIEF OVERVIEW ON HORMONES IN ANIMALS

Hormones are chemical compounds that are secreted directly into the bloodstream by various endocrine glands. They are then transported to specific organs, where they regulate various metabolic processes. The target organ is the specific organ that is influenced by the hormone. Specialized receptors enable target cells to recognize and respond to specific hormones (Squires, 2024). Glands in animals' bodies can be classified into two distinct types.

- Endocrine glands, such as the pituitary and adrenal glands, lack ducts and instead release their secretions directly into the bloodstream, delivering them to the specific site of action.
- Exocrine glands, which are responsible for transporting sweat, liver, and other secretions through ducts.

Below is a list of the different hormones, their functions, and the corresponding glands.

- The hypothalamus gland secretes gonadotropin-releasing hormone (GnRH), which stimulates the release of gonadotropins from the pituitary gland.
- The pituitary gland regulates body growth through the secretion of growth hormones.

- Gland Thyroid hormone, specifically thyroxine, regulates the metabolic processes in the body.
- The pancreatic gland regulates blood sugar levels through the secretion of glucagon and insulin.
- The adrenal gland regulates blood pressure and heart rate through the release of epinephrine.
- Testes are responsible for sperm development and the development of masculine traits.
- Ovaries are responsible for the production of two important hormones, progesterone and estrogen. These hormones play a crucial role in the development of eggs and the manifestation of feminine characteristics (Squires, 2024).

Hormones regulate the function of every organ. Squires (2024) states that they exert influence on multiple facets of sexual characteristics, reproduction, and growth and development. Even small doses of hormones can elicit significant and easily observable reactions in the body. Hormones are predominantly composed of steroids or proteins. Having an excessive or insufficient amount of something is always precarious. Hormonal imbalance refers to the condition where there is an inadequate or excessive production of one hormone by another hormone (Squires, 2024).

7.1.1. Categories of Hormonal Imbalances

- **Adrenal Insufficiency:** This is also known as Addison's disease, is characterized by inadequate secretion of adrenaline. Addison's disease is characterized by symptoms such as fatigue, dehydration, and abnormalities in the skin.
- **Cushing's Disease:** This is caused by excessive secretion of ACTH from the pituitary gland, leading to an overactive adrenal gland. Squires (2024) identifies this as a variant of Cushing's syndrome that specifically impacts individuals with heightened levels of corticosteroids.
- **Acromegaly, or Gigantism:** This is a medical condition characterized by excessive growth in children as a result of the pituitary gland's overproduction of growth hormone.
- **Hyperthyroidism:** This is a condition where the thyroid gland produces an excessive amount of thyroxine hormone. Common symptoms of this condition include perspiration and an accelerated heart rate.

- **Hypothyroidism:** This condition occurs when the thyroid gland fails to produce adequate levels of thyroxine. Some symptoms of goiter, which is the enlargement of the thyroid gland, include exhaustion, constipation, and dry skin.

Hormones in animals can be classified based on their chemical structure and function. Hormones can be classified into different categories based on their chemical composition.

1. **Peptide/Protein Hormones:**These hormones consist of interconnected polypeptide chains made up of amino acids. The secretory vesicles serve as the site for synthesis and storage of peptide hormones. They are situated within the cellular membrane and are released through exocytosis, enabling them to be liberated from the original cell. Instances of hormones, such as oxytocin, vasopressin, and ADH (antidiuretic hormone), are produced in the brain and released into the bloodstream by the posterior pituitary gland (Squires, 2024).

This category encompasses large glycoproteins such as FSH (follicle-stimulating hormone), which is also synthesized by the pituitary gland, as well as small proteins like growth hormones, which are also produced by the pituitary gland. After the vesicles are filled, a stimulus triggers a reaction that leads to the release of the contents, such as the secretion of insulin in response to high blood glucose levels. These hormones are hydrophilic and hydrophobic. Due to their inability to traverse the plasma membrane of cells, the receptors are situated on the exterior surface of the target cells (Squires, 2024).

2. **Hormones in Steroids**: These hormones are derived from lipids and are produced from cholesterol. The precursors produce them as needed and they leave the parent cell through simple diffusion. These hormones bind to proteins in the bloodstream and usually cause the desired outcome of stimulating the production of new proteins (Squires, 2024). These hormones are the main category of lipid hormones. Typically, they are chemical compounds that belong to the class of alcohols or ketones. For example, estrogen and testosterone are hormones produced by the female and male reproductive systems, respectively. These hormones are not soluble in water. Unlike peptide hormones, steroid hormones have a longer duration of action because they bind to transport proteins in the bloodstream. The hormone cortisol. The half-life of the substance varies between 60 and 90 minutes.

3. **Hormones Derived from Amino Acids:** These molecules are derived from the amino acids tyrosine and tryptophan and are comparatively

smaller in size. Catecholamines, namely norepinephrine and epinephrine, are secreted by the adrenal gland. The pineal gland, situated in the brain, governs the circadian rhythm by releasing melatonin.

4. **Female Hormones:**This, also known as estrogen and progesterone, are responsible for the development of secondary sexual characteristics in women. A minority of them are also generated spontaneously in males, although they do not possess dominance. Due to their greater prominence within the female body, these factors play a significant role in a woman's fertility.

The female hormones include the following:

- **Progesterone:** This is a hormone that aids in conception, prepares the lining of the uterus for the fertilized egg, and decreases estrogen production following ovulation.
- **Estrogen:** It plays a vital role in the processes of menstruation, menopause, puberty, and pregnancy. HCG is secreted by the placenta.
- **Testosterone:** Typically associated with males, is also produced in small amounts by women, resulting in levels similar to estrogen in males (Squires, 2024).

7.2. THE PROCESS OF PRODUCING AND CONTROLLING HORMONES

Hormone synthesis and regulation are essential for maintaining a wide range of bodily functions. This section primarily examines three key regulatory processes: stress, metamorphosis, and blood glucose. Although invertebrates are occasionally mentioned, the primary emphasis is on vertebrates. Regulation of blood glucose is crucial for the optimal functioning of animal cells. Glucose serves as the primary energy substrate for the majority of animal cells, and it is distributed throughout the body via the circulatory system. The ideal blood glucose concentration for humans is approximately 90 mg/100 mL of blood. Following a meal, carbohydrates undergo the process of digestion and are subsequently assimilated into the circulatory system. The body typically demands a lower amount of glucose than the current quantity available, thus the surplus glucose must be expelled and stored for future utilization. However, while fasting, the body must sustain blood glucose levels in order to maintain adequate energy supply.

The pancreas has a vital function in controlling blood glucose levels through the secretion of two hormones: insulin and glucagon. Insulin is secreted by the beta cells of the pancreas in response to elevated blood glucose levels, such as those that occur after a meal. Insulin exerts multiple effects that aid in reducing blood glucose levels. It accelerates the uptake and utilization of glucose by target cells, stimulates the liver to convert glucose into glycogen for storage, improves the transportation of glucose into specific cells such as muscle and liver cells, and facilitates the conversion of glucose into fat in adipocytes. The combined impact of these effects results in a decrease in blood glucose levels, which is referred to as the hypoglycemia "low sugar" effect. Insulin production is controlled by a negative feedback mechanism, in which increasing levels of blood glucose suppress the synthesis of additional insulin.

Conversely, the pancreatic alpha cells release glucagon when blood glucose levels decrease below the standard range, such as during periods of fasting or following intense physical exertion. Glucagon exerts a contrasting influence to insulin by elevating blood glucose levels. It activates the liver to metabolize glycogen and secrete glucose into the bloodstream. Glucagon additionally stimulates the transformation of amino acids from the bloodstream into glucose and triggers the release of fatty acids into the bloodstream from adipose cells. These processes elevate blood glucose levels and contribute to the maintenance of homeostasis. Glucagon production, like insulin, is controlled by a negative feedback mechanism, in which increasing levels of blood glucose suppress the production of additional glucagon.

Insulin and glucagon collaborate to regulate blood glucose levels within a homeostatic range. Glucagon is released when blood glucose levels decrease, stimulating the liver to degrade glycogen and release glucose into the bloodstream, leading to an elevation in blood glucose levels. In contrast, an increase in blood glucose levels triggers the release of insulin, which prompts target cells to uptake glucose and the liver to convert glucose into glycogen, resulting in a reduction of blood glucose levels.

Elevated blood glucose levels that are not regulated can result in severe medical conditions, such as diabetes mellitus. Diabetes mellitus can arise from impaired insulin function, which can be caused by reduced sensitivity of tissue cells to insulin or inadequate production of insulin by the pancreatic beta cells. This leads to the condition known as hyperglycemia, characterized by elevated blood glucose levels and impaired glucose uptake by cells. Elevated and sustained levels of glucose in the blood can result in harm to peripheral body tissues, eyes, kidneys, and the cardiovascular system as a consequence of nerve damage.

The thyroid gland is essential for the regulation of growth and metabolism in vertebrates. Thyroxine (T4) and triiodothyronine (T3) are hormones that play a crucial role in initiating the development of new structures, rearranging internal organ systems, and regulating various processes during metamorphosis in species like amphibians (Mullur et al., 2014). Metamorphosis in insects is regulated by a distinct set of hormones. The corpus allatum in the brain secretes the juvenile hormone, which maintains the insect in its larval state. The prothoracicotropic hormone, secreted by a different endocrine gland located in the brain, directs the prothoracic gland to secrete the hormone ecdysone. Ecdysone's role in promoting either metamorphosis or molting is contingent upon the presence or absence of juvenile hormone. Molting to the next stage is triggered by elevated levels of ecdysone and juvenile hormone, whereas low levels of these hormones induce metamorphosis into an adult (Mullur et al., 2014). The primary role of endocrine hormones is to maintain homeostasis, and stressors have the potential to disturb this balance. Immediate stressors elicit the "fight-or-flight" stress response, which raises blood glucose levels and supplies energy for immediate physical reactions to stress. This physiological response, known as the "fight-or-flight" response, is a universal characteristic found in all vertebrates. It serves to prepare the body for survival by enabling actions such as fighting, fleeing, or freezing in response to a perceived threat (Mullur et al., 2014).

Nevertheless, certain stressors, such as illness or physical harm, can endure for an extended duration. Glycogen stores provide for immediate energy requirements; however, they are finite and inadequate for prolonged energy demands. Glucocorticoids are essential in meeting long-term energy requirements in such situations. Glucocorticoids stimulate the production of new glucose, release fat and protein reserves, store glucose for the brain, and preserve salt and water (Mullur et al., 2014). The sympathetic nervous system regulates the stress response via the hypothalamus. The adrenal cortex, responsible for regulating long-term stress responses, receives signals from the hypothalamus through the secretion of adrenocorticotropic hormone (ACTH) by the anterior pituitary gland. The adrenal medulla, responsible for short-term stress responses, receives neural signals from the brain (Mullur et al., 2014). When faced with stressful situations, the body secretes hormones that rapidly increase energy levels. The adrenal medulla secretes norepinephrine and adrenaline, which stimulate the hydrolysis of glycogen into glucose and the liberation of glucose from muscle and liver cells. Additionally, these hormones elevate blood pressure, respiratory rate, and metabolic rate, while also modifying blood flow patterns to redirect blood towards the heart, brain, and skeletal muscles (Mullur et al., 2014).

Nevertheless, the immediate stress reaction controlled by adrenaline and norepinephrine is not viable for extended durations. Additional hormones are implicated in such instances. When the hypothalamus triggers a long-lasting stress response, the anterior pituitary gland secretes adrenocorticotropic hormone (ACTH). ACTH stimulates the adrenal cortex to secrete corticosteroids, which are a type of steroid hormone. There are two primary categories of corticosteroids: mineralocorticoids, such as aldosterone, and glucocorticoids, such as cortisol (Mullur et al., 2014). Glucocorticoids are essential for the body's response to prolonged stress. They enhance the production of glucose from lipids and proteins, resulting in an increase in blood glucose levels. Glucocorticoids facilitate the degradation of fat into fatty acids within adipose tissue, which are subsequently released into the bloodstream to generate ATP. In addition, glucocorticoids inhibit the immune system in order to preserve energy (Mullur et al., 2014). Conversely, mineralocorticoids, particularly aldosterone, elevate blood pressure and volume by promoting the kidneys to retain water and salt ions (Mullur et al., 2014).

7.3. HORMONE SIGNALING PATHWAYS

Animal hormone signaling pathways create a complex network of communication that synchronizes vital functions in living organisms. Hormones, which are bioactive substances synthesized in one region of the body but exert their effects on remote target cells, play a role in these pathways. Hormones are able to connect endocrine cells to their intended targets by traveling through the bloodstream. When a hormone binds to its specific receptor on a cell, it can either enhance or inhibit normal physiological processes. This binding activates complex signal transduction systems, which ultimately lead to cell-specific responses (Larigot et al., 2018).

7.3.1. Chemical Communication

In the realm of living cells, there exist two distinct forms of communication. Intercellular signaling refers to the process of communication between different cells, while intracellular signaling refers to the process of communication within a single cell (Larigot et al., 2018). Familiarity with the Latin origins of the prefixes inter- and intra-facilitates the easy recollection of their respective meanings. Inter- signifies "between" as seen in instances such as intersecting lines, while intra- denotes "inside" as exemplified by intravenous.

Signaling cells secrete chemical signals in the form of ligands, which are small, usually volatile or soluble molecules. A ligand is a specialized molecule that binds to another molecule and, sometimes, transmits a signal during the binding process. Therefore, ligands can be regarded as signaling molecules. Target cells are the cells that are affected by chemical signals. Ligands bind to the proteins of these cells, which are also referred to as receptors. Various ligands and receptors exist, with each ligand typically binding exclusively to its unique receptor (Larigot et al., 2018).

Animals demonstrate four distinct forms of chemical signaling:

- Direct signaling through gap junctions;
- Autocrine signaling;
- Endocrine signaling; and
- Paracrine signaling.

The primary differentiating factor among the various signaling categories is the distance that a signal must travel within an organism to reach its target cell. Signals elicit varying responses in different cells (Larigot et al., 2018).

In chemical signaling, a cell has the ability to target various types of cells, including itself (autocrine signaling), a cell connected by gap junctions, a neighboring cell (paracrine signaling), or a distant cell (endocrine signaling). Endocrine signaling involves the transportation of ligands through the circulatory system, autocrine signaling occurs within the signaling cell itself, and paracrine signaling affects nearby cells. Gap junction signaling involves the direct transfer of signaling molecules between adjacent cells.

7.3.2. Paracrine Signaling

Paracrine signals are localized signals that specifically act between neighboring cells. Diffusion is the process by which paracrine signals propagate through the extracellular matrix. These signals usually lead to immediate responses that are not long-lasting. Paracrine ligand molecules are typically swiftly degraded by enzymes or cleared by neighboring cells to ensure localized confinement of the reaction. If the signals are released again, removing them will restore the concentration gradient of the signal, allowing it to quickly diffuse into the intracellular space (Larigot et al., 2018).

7.3.3. Endocrine Signaling

This refers to the process by which hormones are released into the bloodstream and travel to target cells or organs to regulate various physiological functions.

Signals emanating from cells located far away are referred to as endocrine signals, and they are produced by endocrine cells. Endocrine glands, such as the pituitary, thyroid, and hypothalamus, contain numerous endocrine cells in the body. Typically, these signals lead to a longer response time but have a more lasting effect. Hormones are signaling molecules that act as ligands in endocrine signaling. They are produced in one part of the body but have a widespread effect on distant regions of the body (Larigot et al., 2018). Hormones are transported throughout the body at a relatively slow pace, primarily through the bloodstream. This allows them to travel long distances between endocrine cells and their target cells. When hormones exert their effects on specific cells, their transportation mechanism causes them to become diluted, resulting in low concentrations. In contrast to paracrine signaling, where ligand concentrations can be significantly elevated in a localized area.

7.3.4. Autocrine Signaling

This refers to a type of cell communication where a cell releases signaling molecules that bind to receptors on its own surface, leading to a response within the same cell. Cells that can both bind to and release a specific molecule are capable of generating autocrine signals. The prefix "auto-" denotes self, serving as a reminder that the signaling cell transmits a signal to itself. This suggests that the signaling cell and the target cell could potentially be identical or highly similar cells. During an organism's early development, it is crucial to ensure that cells differentiate into the correct tissues and perform their intended functions. This process of communication is a common occurrence.

Furthermore, autocrine signaling regulates both the inflammatory response and the perception of pain. Furthermore, a cell that is infected with a virus possesses the capability to initiate programmed cell death, thereby eradicating the virus from the infected cell. On certain occasions, the liberated ligand also impacts adjacent cells of the identical kind. To ensure the desired developmental outcome during embryonic development, the process of activating a cluster of adjacent cells can help guide the transformation of identical cells into the same type of cell (Larigot et al., 2018).

7.3.5. Direct Signaling Through Gap Junctions

Gap junctions refer to the connections between adjacent cell plasma membranes in animals, while plasmodesmata are the equivalent structures in plants. Intracellular mediators, which are small signaling molecules, can traverse between the two cells via these water-filled channels. Macromolecules such as proteins and DNA are too big to pass through the channels, whereas small molecules like calcium ions (Ca^{2+}) can move between cells. By maintaining cellular autonomy, the specialization of channels enables the rapid and uncomplicated transmission of signals. For a group of cells to synchronize their response to a signal, it is necessary to transfer signaling molecules. This allows the current state of the cell directly next to the target cell to be communicated (Larigot et al., 2018).

7.4. ENDOCRINE SYSTEM IN ANIMALS

The endocrine system in animals is essential for regulating bodily functions by producing, secreting, and controlling hormones. Hormones function as chemical messengers that influence the functioning of organs and cells, contributing to the maintenance of homeostasis in the body. The endocrine system exerts influence over multiple facets of an organism's existence, encompassing development, sexuality, growth, and metabolism. Endocrine system disorders, such as diabetes mellitus and thyroid disease, are prevalent among humans (Kumar et al., 2022).

The endocrine system also has a substantial impact on the metamorphic process of animals that experience it. As an illustration, the transformation from tadpole to frog is a multifaceted and precise procedure governed by the endocrine system, enabling animals to adjust to particular environments and ecological circumstances.

The hormones present in the animal body can be categorized into three distinct types according to their chemical composition: lipid-derived hormones, amino acid-derived hormones, and peptide hormones. Lipid-derived hormones, also referred to as lipid-soluble hormones, have the ability to pass through plasma membranes, unlike amino acid-derived and peptide hormones (Kumar et al., 2022).

Lipid-derived hormones, such as steroid hormones, share a similar structure with cholesterol as they are derived from it. Steroid hormones such as estradiol, which is an estrogen and the primary female sex hormone, and testosterone, which is an androgen and the main male sex hormone, serve as examples. These hormones are secreted by the male and female reproductive systems, respectively. In addition, the adrenal glands also release other steroid hormones such as cortisol and aldosterone. Lipid hormones exhibit solubility in water and are conveyed in the bloodstream through transport proteins, thereby enabling them to have a longer duration of presence in the body compared to peptide hormones (Kumar et al., 2022).

Amino acid-derived hormones are small molecules that are derived from the amino acids tyrosine and tryptophan. Some examples of these hormones are epinephrine, norepinephrine, and thyroxine. The adrenal glands' medulla produces epinephrine and norepinephrine, which play a role in the fight-or-flight response. The thyroid gland produces thyroxine, while the pineal gland in the brain produces melatonin, which is responsible for regulating sleep cycles.

Peptide hormones consist of polypeptide chains comprising amino acids. These include small polypeptide chains such as oxytocin and antidiuretic hormone. They are synthesized in the brain and secreted into the bloodstream by the posterior pituitary gland. Peptide hormones encompass not only small proteins such as growth hormones, but also large glycoproteins like follicle-stimulating hormone. These hormones are synthesized by the pituitary gland. Peptide hormones, like insulin, are stored in vesicles inside the cells that produce them and are released when there is a stimulus, such as high blood glucose levels. The solubility of these hormones is limited to water and they are unable to pass through plasma barriers due to their insolubility in fat (Kumar et al., 2022).

Hormones exert their effects through the process of binding to receptors located on target cells. This interaction has the potential to induce alterations in the quantity of receptors present on the target cell. Hormones can exert their influence on cells through two mechanisms: either by binding to intracellular hormone receptors or by binding to hormone receptors located on the plasma membrane. These effects can be either direct or indirect in nature. Lipid-soluble hormones have the ability to enter cells by passing through the plasma membrane and binding to DNA in order to regulate gene transcription (Kumar et al., 2022).

The endocrine system is comprised of multiple endocrine glands that secrete hormones to facilitate communication and regulation of bodily functions. These hormones control the processes of development, growth, energy metabolism, reproduction, and behaviors. The nervous system regulates the activities of endocrine, muscular, and neuronal cells by releasing neurotransmitters and neurohormones (Kumar et al., 2022).

7.5. ANIMAL HORMONE DYSFUNCTION DISORDERS

Endocrine disorders are common in animals. Companion animals, such as dogs or cats, are more likely to have their medical conditions diagnosed and treated compared to larger animal species. The clinical signs and therapy vary significantly depending on the specific organ system affected and whether the cause is an excess or deficiency of hormones. Multiple factors can contribute to the development of endocrine disorders. Excessive or inadequate production of hormones, malfunctioning receptors, and interference with the normal hormone removal processes can occur. Hormonal imbalances can occur due to various factors (Kooistra et al., 2009).

The primary causes of excessive hormone production in veterinary species are either the presence of tumors in secondary locations that stimulate the overproduction of hormones (such as pituitary-dependent hyperadrenocorticism, also known as Cushing's disease, in dogs) or the presence of adenomas or hyperplasia in the endocrine tissue itself (such as adrenal-dependent hyperadrenocorticism in dogs or hyperthyroidism in cats). Hormone deficit syndromes commonly arise due to an autoimmune reaction that leads to the destruction of an endocrine organ (Kooistra et al., 2009). Clinical symptoms resulting from an endocrine tissue malfunction can occur due to a problem originating from the hormone's main source or from a disturbance that affects the hormone's secretion or action in another location (Kooistra et al., 2009).

7.5.1. Pathogenesis of Endocrine Diseases in Animals

Abnormal hormone release often occurs due to irregular endocrine tissue, which not only produces excessive amounts of hormone but also fails to respond appropriately to feedback signals. An excess of stimulation can lead to the excessive production of hormones by endocrine tissue. For instance, renal illness can result in parathyroid hyperplasia and the excessive secretion of parathyroid hormone (PTH).

Specific types of kidney disease can result in hyperphosphatemia. It leads to a decrease in the synthesis of calcitriol, also known as 1,25-dihydroxycholecalciferol, which is the biologically active form of vitamin D. Decreased levels of calcitriol result in decreased levels of calcium in the extracellular fluid, which subsequently stimulates the release of parathyroid hormone (PTH) (Kooistra et al., 2009).

Nonendocrine tissues have the ability to produce and release hormones in sufficient amounts to cause symptoms. For example, certain types of tumors, like lymphomas and apocrine gland tumors of the anal sac in dogs, have the ability to generate PTH-related protein (PTHrP). This protein can imitate the actions of PTH and result in hypercalcemia, a paraneoplastic condition. Various syndromes have been found to be associated with hormone secretion deficiencies and absences. The prevailing belief is that the cause is the autoimmune destruction of endocrine tissue by cells. Endocrine hypofunctions such as dog hypothyroidism, type I diabetes mellitus, primary hypoparathyroidism, and primary hypoadrenocorticism (Addison disease) are caused by the loss of primary tissue (Kooistra et al., 2009).

Feedback pathways are activated in response to early tissue loss, leading to an increase in the activity of the remaining tissue and the production of hormones. In cases of Addison's disease, the disappearance of the adrenal cortex leads to an increase in pituitary ACTH production. The remaining tissue becomes fully stimulated due to the heightened trophic support, often leading to the secretion of sufficient hormones that delay the occurrence of deficiencies until the loss of tissue eliminates the hormone supply. Disruption in tissues located far from the source of the hormone can also lead to disorders that manifest as clinical indications of reduced endocrine activity. A deficiency in pituitary thyroid-stimulating hormone (TSH) reduces the level of stimulation needed by the thyroid gland to produce and release thyroxine (T4) and triiodothyronine (T3), resulting in secondary hypothyroidism.

Administration of glucocorticoids can lead to the atrophy of the cortisol-producing zones in the adrenal cortex (Kooistra et al., 2009). The exogenous steroid induces negative feedback in the pituitary gland, leading to the suppression of ACTH release and resulting in adrenal cortical atrophy.

Endocrine hypofunction can also occur due to tissue loss caused by the compressive or destructive growth of nonfunctioning tumors. Alterations in the tissue's responsiveness to hormones, such as insulin resistance (as observed in type II diabetes), can also result in endocrine disorders.

The cause of nephrogenic diabetes insipidus is the renal insensitivity to the activities of vasopressin, also known as antidiuretic hormone. The presence of congenital abnormalities in the vasopressin receptor may be associated with the inability of the kidneys to respond to vasopressin. However, this lack of sensitivity is often a result of other medical conditions such as pyometra or hyperadrenocorticism (Cushing syndrome), or abnormal levels of ions such as hypokalemia or hypercalcemia.

7.5.2. Manifestations of Endocrine Disorders in Animals

The clinical manifestations of endocrine disorders in animals can exhibit variability based on the particular pathology. Cats suffering from hyperthyroidism often experience weight loss along with an increased appetite, a condition known as polyphagia. Conversely, dogs suffering from hypothyroidism may encounter symptoms such as increased body weight, skin issues, and lack of energy. Dogs suffering from hyperadrenocorticism, also known as Cushing syndrome, frequently experience alopecia, as well as heightened hunger, urination, and thirst (polyuria and polydipsia). Dogs with typical hypoadrenocorticism (Addison disease) exhibit nonspecific symptoms such as lethargy, loss of appetite, and weight loss (Kooistra et al., 2009).

A thorough physical examination and detailed medical history are crucial when diagnosing endocrine disorders in animals. The initial assessment involves conducting a comprehensive blood test, including a complete blood count (CBC) and biochemical analysis. The process of differential diagnosis involves analyzing examination findings and test results. In some cases, specific hormone tests, such as those for thyroid and adrenal function, may be performed to either confirm or rule out certain diagnoses (Kooistra et al., 2009).

Endocrine diseases in animals can be treated through various methods, such as surgical interventions, medical management, radiation therapy, nutritional therapy, and hormone replacement therapy. Hormone replacement therapy, such as administering insulin for diabetes mellitus or providing thyroid hormone replacement therapy for hypothyroidism, can be utilized to rectify hormone insufficiencies. Nevertheless, the treatment of diseases caused by protein and polypeptide hormone deficiencies can be difficult due to the high cost and limited accessibility of the hormone. Treatment options for endocrine disorders characterized by hyperactivity, such as feline hyperthyroidism, include surgical procedures, radiation therapy, medication, and dietary changes. The selection of treatment is contingent upon the particular disorder (Kooistra et al., 2009).

Treating deficiencies in protein and polypeptide hormones using hormone replacement therapy can be challenging. It is possible that the specific form of the hormone for a particular species may not be accessible, which would necessitate administering multiple injections each day or taking into account the potential for developing antibodies and experiencing anaphylaxis. Thyroid hormone and steroids are commonly ingested orally, whereas specific protein and polypeptide hormones or their analogs can be administered through alternative routes besides injection. It is crucial to monitor the clinical response and perform therapeutic blood monitoring in order to assess the efficacy of hormone replacement therapy (Kooistra et al., 2009).

After the surgical removal of an endocrine tumor, hormone replacement therapy may be required temporarily. Nevertheless, the healthy tissue that remained after the disease underwent atrophy typically recovers its functionality swiftly, obviating the necessity for continuous administration of hormone replacement medication. Individualized dosage regimens are necessary for patients due to the variability in drug bioavailability in animals. Glucocorticoids are frequently prescribed drugs for endocrine disorders, primarily due to their ability to reduce inflammation and alleviate allergies. It is essential to comprehend the potential side effects, such as the occurrence of hyperadrenocorticism, in order to use them appropriately. Intermittent treatment regimens can mitigate the adverse effects of prolonged therapy.

7.6. ROLE OF HORMONES IN ANIMAL HEALTH

Hormones are essential for maintaining animal health as they control and regulate a wide range of physiological processes in their bodies. Hormones are bioactive substances secreted by specialized glands, which are then transported through the bloodstream to specific target cells or organs. Once they reach their destination, hormones initiate specific physiological responses. Hormones play a role in regulating homeostasis, controlling growth and development, managing reproduction, regulating metabolism, influencing behavior, and modulating immune function. Disturbances in hormonal equilibrium can result in diseases in animals, underscoring the significance of comprehending these intricate hormonal interconnections for their welfare (Nauck & Meier, 2018).

Hormones exert a profound influence on animal behavior through complex interactions. They exert a direct impact on neuronal activity by means of neural modulation, thereby modifying the functioning of the brain and influencing an animal's response to threats or challenges. Hormones exert regulatory control over gene expression, either by stimulating or inhibiting specific genes, thereby influencing behavior in the long run. In addition, hormones and neurotransmitters interact within neurotransmitter systems in the brain, where hormone-regulated serotonin levels influence both aggression and mood. Recent studies have shown that specific steroids have the ability to rapidly control stress responses and social behaviors within a matter of seconds or minutes. This is achieved through nongenomic mechanisms that involve receptors on the cell membrane (Nauck & Meier, 2018).

Gaining a comprehensive understanding of the influence of hormones on animal behavior is crucial not only for their general well-being but also for gaining valuable insights into human health. Several hormones that influence animal behavior are also found in humans, providing novel insights into stress-related disorders, reproductive problems, and atypical behavior. Through the examination of hormonal regulation in animal behavior, scientists can enhance their comprehension of these intricate processes and potentially create novel therapies for corresponding human conditions.

7.7. PROSPECTS FOR ADVANCEMENTS IN HORMONE RESEARCH

The existence of growth rate disparities between sexes that are influenced by hormones has been acknowledged for a considerable period of time. Furthermore, it has been established that intact males exhibit a higher rate of growth and feed conversion efficiency (FCE) compared to castrate males. Therefore, it was logical to conduct research that aimed to utilize hormones and other substances that naturally occur or are artificially created and have hormonal effects, in order to enhance production. Diethylstilbestrol (DES) and hexoestrol (HEXO) were administered to cattle more frequently in the United States and the United Kingdom, beginning in the mid-1950s. These substances were either implanted or added to their feed as additives. Additional types of drugs gradually became readily available. In general, this type of treatment has enhanced the quality of the carcass by increasing the ratio of lean to fat. It has also improved feed conversion efficiency (FCE) and increased daily weight gains by 10% to 15% (Schug et al., 2016). It is crucial to differentiate between manufactured or other compounds, which the body may lack the necessary enzymes to metabolize, and hormones themselves, whose metabolism is well understood in the body. The assertion that zero-tolerance residue levels are insignificant becomes invalid when natural hormones are employed in animal production, as these substances are detectable and exhibit significant variability in their quantities within the bodily fluids and tissues of all animals, regardless of treatment). However, the scenario varies for other medications that contain a hormone component. However, it is logical to weigh the potential risks against the undeniable advantages that certain chemicals offer in terms of producing animal protein, especially when the levels of these chemicals are extremely low (Schug et al., 2016). Considering these perspectives, the significance of hormones in animal husbandry may seem trivial, prompting the question of the importance of prioritizing various initiatives aimed at enhancing output and productivity. Nevertheless, it is currently impractical to adopt a single approach on a worldwide scale while disregarding all others. Preparations that offer both positive effects on crop yield and efficient use of feed, while posing minimal risk to consumers, will consistently be in demand. Moreover, in the future, the use of hormonally active drugs may not be limited to those that are currently available. According to Schug et al. (2016), all existing substances, regardless of whether they are natural or synthetic, possess the characteristic of being resistant to decomposition by the body.

Protein hormones, in contrast, exhibit a fundamentally distinct behavior as they are fully degraded into amino acids and do not leave any detectable residue. One example of such a hormone is the growth hormone, which stimulates both growth and milk secretion, even in cows with high milk production. Currently, only a restricted quantity of this anabolic hormone is available for research purposes. Recent advancements in the utilization of recombinant DNA technology have made it feasible to produce species-specific peptide hormones on a large scale through microbial manufacturing. The combination of advanced hormone usage in animal production and the development of smaller automated administration devices for subcutaneous application may usher in a new era (Schug et al., 2016).

Chapter **8**

Animal Nutrient Metabolism

Contents

Upon consumption of food, the fascinating process of animal nutrient metabolism commences. Animals, like humans, acquire their sustenance from food, which consists of vitamins, minerals, proteins, fats, and carbohydrates. Enzymes break down these nutrients into simpler forms during the process of digestion that takes place after ingestion. Subsequently, these modified nutrients can be absorbed into the bloodstream and distributed to various tissues and organs. Ultimately, these nutrients undergo conversion into energy, serve as building blocks for growth, and function as essential molecules for sustaining life through metabolism—the intricate interplay of chemical reactions. Every heartbeat and movement in the animal kingdom is fueled by nutritional metabolism, regardless of whether it is a hummingbird feeding on nectar or a grazing cow. This chapter will explore different aspects of nutrient metabolism in animals, such as an overview of nutrients in animals, the role of vitamins and minerals in animal nutrition, the process of digestion and absorption of nutrients, the metabolism of essential nutrients, the nutrient requirements in animals, disorders related to nutrient metabolism in animals, and finally, the significance of nutrients in animal health.

8.1. A BRIEF OVERVIEW OF NUTRIENTS IN ANIMALS

Animals rely on nutrients from external sources and lack the ability to synthesize their own food, which is in contrast to plants. These organisms rely on plants, either directly or indirectly (for instance, by consuming animals that feed on plants), while some organisms depend on both. Every organism, including humans and animals, depend on food as a source of energy. Ruminants (2007) emphasized the importance of developing, repairing, and functioning.

Animals generally have more complex and diverse dietary needs compared to plants, which have simpler nutritional requirements. Animals ingest lipids, proteins, carbohydrates, minerals, and vitamins, along with other essential nutrients.

- All animals primarily derive their energy from carbohydrates. In contrast to plants, which synthesize carbohydrates through photosynthesis, animals obtain their carbohydrates from the external environment. According to Ruminants (2007), carbohydrates account for approximately 50% to 66% of the total daily calorie intake of an animal. Glucose is the most prevalent carbohydrate used for energy. Some of the energy produced by the metabolic process of this monosaccharide in cellular respiration is used to generate adenosine triphosphate (ATP). Starch, sucrose, lactose, and maltose are additional beneficial carbohydrates.

- Lipids form the protective coverings around nerve fibers, organelles, cellular membranes, and certain hormones. Fats, a type of lipid, are highly efficient sources of energy.
- Nucleic acids, such as deoxyribonucleic acid (DNA), ribonucleic acid (RNA), and ATP, are all constituents of DNA, RNA, and ATP. Animals obtain nucleic acids from animal and plant tissues, especially those containing nuclei. Nucleotides are absorbed by cells after nucleic acids are hydrolyzed during digestion.
- The mammalian body is composed of proteins. Proteins are essential for the functioning of membranes, organelles, and the cytoplasm (Ruminants, 2007). In addition to being essential components of enzymes, amino acids are also the fundamental building blocks of tendons, ligaments, and muscles. Proteins are composed of a total of 20 distinct amino acids. Although the body can synthesize many amino acids, a significant portion must be acquired through dietary sources. During the process of digestion, proteins are hydrolyzed into their individual amino acids, which are subsequently absorbed by the body (Ruminants, 2007).
- Animals require essential minerals such as magnesium, zinc, sulfur, potassium, and phosphorus. Usually, animals obtain these nutrients from plants by consuming them as part of their diet. Vitamins are essential micronutrients required for optimal animal health. Vitamins have the ability to dissolve in either fat or water. While fat-soluble vitamins are stored in lipid droplets in the liver, water-soluble vitamins require frequent intake. Several essential vitamins include vitamin A, which promotes good eyesight, vitamin B, which is involved in cellular respiration through molecules such as FAD, NAD, and coenzyme A, and vitamin D, which aids in the absorption of calcium in the body (Ruminants, 2007).

Animals employ a diverse array of feeding strategies to acquire the necessary nutrients. Sponges consume food particles that pass through their pores (Ruminants, 2007). Food becomes ensnared on the adhesive surfaces of the appendages of various marine creatures, such as sea cucumbers. Clams and oysters are mollusks that acquire nutrients by filtering particles through a mucous layer in their gills. Certain arthropods exclusively ingest fluids (Ruminants, 2007).

Some animals possess specialized organs for the purpose of grasping, chewing, and ingesting food due to their reliance on food masses (Ruminants,

2007). Herbivores exclusively consume plants, while carnivores exclusively consume other organisms. Humans are classified as omnivores, which indicates that they have a diet that includes both plant and animal matter. Animals can be classified into distinct groups according to their feeding habits (Figure 8.1):

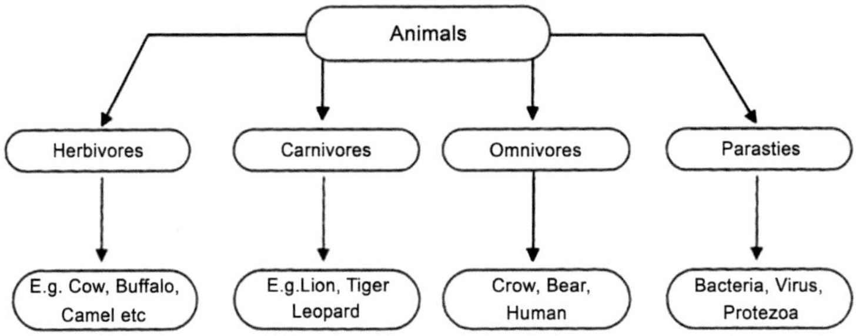

Figure 8.1.*Classification of animals on the basis of food habits.*

Source: Studyadda (2020).

- Herbivorous animals depend on fruits and plants for their sustenance. Herbivores such as sheep, goats, cows, buffaloes, and others are classified as ruminants.
- Carnivores are organisms that depend on other organisms for their sustenance. Examples of carnivores are wolves, lions, and tigers (Ruminants, 2007).
- Omnivores are organisms that consume both vegetation and animals. Omnivores encompass a variety of species such as dogs, ravens, bears, and humans.

8.1.1. Various Animal Nutritional Classifications

Animals have the ability to consume a diverse range of diets, which may include:

- Filter feeding refers to the process of obtaining nutrients from suspended particles in water. Fish utilize it frequently.
- Deposit feeding is the act of extracting nutrients from particles suspended in soil. Earthworms absorb food through a process known as ingestion.
- Fluid feeding refers to the process of obtaining nutrients by consuming the bodily fluids of other organisms. Mosquitoes and honey bees exemplify this method of food intake.
- The act of consuming an entire organism to obtain nutrients is referred to as bulk feeding. Python is an exemplar.

- Ram feeding and suction feeding both involve the ingestion of prey by utilizing the surrounding bodily fluids. This swallowing technique is commonly observed in aquatic predators, such as bony fish.

8.1.2. Process of Animal Nutrition

The animal's nourishment process involves the following steps:

- **Ingestion:** This refers to the process of consuming food.
- **Digestion:** During this process, the larger food particles are broken down into smaller particles that are soluble in water. According to Ruminants (2007), food can undergo chemical or physical breakdown.
- **Absorption:** After the process of digestion, the nutrients from the food are taken up by the bloodstream through the intestinal wall.
- **Assimilation:** The human body utilizes the nutrients obtained from digested food for the purpose of energy production, cellular growth, and tissue repair.
- **Egestion:** This refers to the process of eliminating waste materials from the body. Undigested food is excreted from the body in the form of feces. This process is referred to as egestion (Ruminants, 2007). Figure 8.2 depicts the overview of animal nutrition.

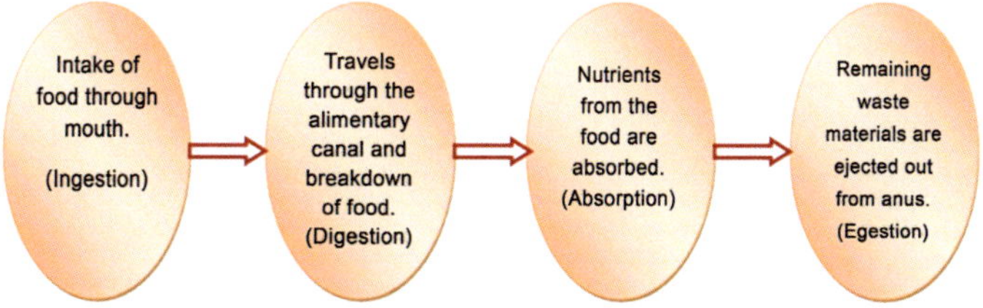

Figure 8.2. *Overview of animal nutrition.*

Source: BYJUS (2023).

8.2. THE ROLE OF VITAMINS AND MINERALS IN ANIMAL NUTRITION

The physiological and metabolic functions of an animal's body rely on minerals, which are inorganic substances. The animal's body weight consists of approximately 4% mineral matter, which is essential for the animal's survival and well-being. Minerals are crucial for all biological processes in the body, more so than any other category of nutrients. The activities encompassed by gene expression and regulation include controlling the activity of enzyme systems and cells, vitamin activity and function, osmotic balance, detoxification, immunity, cell membrane function, acid-base balance and regulation, and growth and structural support (e.g., bone) (Saha & Pathak, 2021).

The classification of minerals into macro and micro (trace) minerals is determined by their dietary requirements rather than their physiological importance. Macro minerals are essential minerals that are required in substantial amounts (> 0.01%) in the diet and are present in the animal body in significant concentrations. Macro minerals, such as calcium, phosphorus, magnesium, sulfur, and electrolytes (sodium, potassium, and chloride), are exemplified by Saha & Pathak (2021).

Microminerals are required in quantities measured in parts per million, milligrams, micrograms, or trace levels, which are less than 0.01%. The microminerals covered include manganese, zinc, iron, copper, selenium, iodine, cobalt, molybdenum, and chromium. Essential nutrients must be supplied to a diet in the form of compounds that are combined with other elements. They cannot be added in their pure elemental forms (such as NaCl, $CaCO_3$, $MnSO_4$, etc.).

8.2.1. Calcium and Phosphorus

Phosphorus (P) and calcium (Ca) both function as structural components in the bodies of animals. Hydroxyapatite, a compound present in bones and teeth, comprises approximately 99% of the calcium and 80% of the phosphorus found in animal organisms. Around 1% of calcium is present in cellular fluids and serves various metabolic and physiological functions, such as blood coagulation, nerve impulse maintenance, cell permeability, activation of specific enzymes, muscular contraction, and the function of ion channel activators (Saha & Pathak, 2021).

Phosphorus, found in the body's soft tissues, is essential for phosphorylation events and cellular oxidative pathways that are involved in energy consumption. Phosphorus is a constituent of adenosine triphosphate (ATP), a molecule that undergoes phosphorylation and plays a vital role in energy metabolism. Phosphorylated pentose sugars are found in both deoxyribonucleic acid (DNA) and ribonucleic acid (RNA). Furthermore, phosphorus is an essential constituent of phospholipids present in cell membranes, playing a crucial role in maintaining cellular fluidity and facilitating the transfer of nutrients.

Phosphorus and calcium are crucial for various metabolic processes that sustain the existence of animals (Saha & Pathak, 2021). Grain cereals are rich in phosphorus. Phytate and phytic acid are alternative terms for bonded forms of phosphorus that are present in cereal grains. The availability of P from these limited sources varies between 20% and 60%. The term "available P" is commonly used to describe the unbound forms of P in the diet of monogastric animals due to their lack of the enzyme phytase, which is responsible for releasing P from its bound form. Ruminant animals produce microbial phytase, an enzyme that can break down and release phosphorus (P).

8.2.2. Magnesium

Magnesium is present in the body as phosphates, which is the third most plentiful element, and as carbonates in bone, liver, and skeletal muscle cells. Magnesium plays a crucial role in the structure of the skeletal system and is essential for the activation of numerous enzymes involved in the breakdown and transportation of phosphates within cells. Magnesium is a positively charged ion that is present in the fluid inside cells and plays a role in the breakdown and utilization of proteins and carbohydrates. Magnesium is a vital element for the transmission of nerve impulses and the contraction of muscles, along with calcium, sodium, and potassium (Saha & Pathak, 2021).

The majority of dietary magnesium is absorbed in the ileum. According to Saha & Pathak (2021), the absorption of Mg does not require a carrier. Vitamin D does not have an impact on the absorption of magnesium. The regulation of magnesium in blood and tissue is still uncertain, and parathyroid hormone (PTH) stimulates the release of magnesium from bone. Nutritional secondary hyperparathyroidism is characterized by a decrease in serum magnesium levels and an increase in urine excretion.

Food sources are rich in magnesium. Grass tetany is a common problem observed in livestock that engage in grazing activities. It is also referred to as "wheat grass poisoning." It is particularly prevalent among animals that feed on native pastures or lush meadows abundant in grain forages during the spring season. It is primarily prevalent in meadows characterized by the presence of wheat grass. It has been linked to a rise in springtime trans aconitate levels in the green pasture. Magnesium (Mg) is sequestered by trans aconitate, leading to magnesium insufficiency (Saha & Pathak, 2021). Symptoms of tetany include muscle spasms, retraction of the head, difficulty walking, convulsions, and heightened sensitivity to touch or sound. Potassium and nitrogen hinder the absorption of magnesium. Pastures that are abundantly fertilized tend to contain elevated levels of nitrogen (N) and potassium (K). Animals that feed on these pastures are more susceptible to grass tetany. Intravenous administration of a magnesium solution is performed, magnesium is obtained from different sources, pastures are rotated, and dry forages are given as treatments (Saha & Pathak, 2021).

8.2.3. Vitamins

Vitamins are a group of organic compounds that possess unique chemical properties and are necessary in small quantities for various physiological processes. The term "vitamin" is derived from "vital amine," which refers to a group of substances that have specific roles in metabolism. Although vitamins are organic substances, they do not play a role in the formation of structural compounds or provide energy like other macronutrients. However, they function as coenzymes or precursors of enzymes in various metabolic processes.

Although certain vitamins can be synthesized by rumen and hindgut bacteria or obtained through sunlight exposure, the majority of vitamins must be obtained by animals through their diet (Saha & Pathak, 2021). The presence of vitamin deficiencies in the diet leads to the occurrence of diseases, reduced productivity, compromised animal welfare, and weakened immune systems in animals that are being raised for food. Dietary vitamin requirements are relatively minimal. In recent years, animal diets have been supplemented with high doses of specific vitamins, such as vitamin E, with the aim of enhancing food quality and strengthening immunity.

Vitamins can be classified into different categories based on their chemical structure and solubility.

- Fat-soluble vitamins, including vitamins A, D, E, and K.
- Soluble in water (such as vitamin C and the B complex group).

There are several vitamins that are classified as fat-soluble, such as vitamins A, D, E, and K. Vitamins are absorbed and digested in conjunction with fat. It is crucial to understand the potential harm that can arise from excessive storage of certain fat-soluble vitamins, such as A and D, in the liver and adipose tissue. It is not required on a daily basis. The deficit advances at a comparatively sluggish rate.

Water-soluble vitamins consist of a collection of nine essential vitamins, which encompass vitamin C and all B vitamins. Vitamins that are soluble in water are eliminated from the body through urination once any excess has been dissolved. These substances have a lower toxicity level and do not necessitate any storage. Except for vitamin B_{12}, it is necessary to consume these vitamins on a daily basis. Cofactors are essential components in various biological processes. Deficiency can manifest quite rapidly.

8.3. NUTRIENT DIGESTION AND ABSORPTION

Animals, being chemoheterotrophs, rely on organic molecules that are already synthesized to obtain both energy and carbon. Consumption is the means by which animals accomplish this, a process often equated with digestion.

Indeed, the process commonly known as "digestion" can be categorized into four discrete phases:

1. **Ingestion:** This refers to the act of consuming food.
2. **Digestion:** The process of breaking down food occurs through both mechanical and chemical means.
3. **Absorption:** the physiological process through which the body assimilates nutrients from ingested food.
4. **Elimination:** the act of disposing waste into the environment.

8.3.1. Distinguishing Between Complete and Incomplete Digestive Systems

Animals possess the ability to perform the process of decomposing the tissues of another organism in order to render them suitable for absorption. This is accomplished through one of two primary types of digestive systems:

1. **Incomplete Digestive Systems:**The digestive system of certain organisms, referred to as an incomplete digestive system or gastrovascular cavity, has a single opening that functions as both the mouth and the anus.

Following ingestion, food particles pass through either the mouth or anus and proceed along a canal lined with cells that secrete digestive enzymes, facilitating the breakdown of the meal. The cells that line the gastrovascular cavity absorb the food particles (Bakke et al., 2010). Once the process of digestion is complete, the remaining material is eliminated from the body either through the mouth or the anus as waste. Species exhibiting radial symmetry and possessing this type of digestive system include Ctenophora (comb jellies) and Cnidaria (coral, jellyfish, and sea anemones.

2. **Complete Digestive Systems:**The presence of distinct openings for the mouth and anus in complete digestive systems, also known as alimentary canals, enables regional specialization and one-way transportation of food.

Once food is ingested, it passes through the stomach or gizzard for the purpose of digestion, then moves into the intestine to absorb nutrients, and ultimately exits through the anus to eliminate residual waste (Bakke et al., 2010). With the exception of Platyhelminthes (flatworms), which possess an insufficient digestive system, nearly all bilaterally symmetric animals possess a fully developed digestive tract.

8.3.2. Ruminant Digestion

Ruminants engage in foregut fermentation. Prior to consumption, food undergoes partial mechanical digestion within the oral cavity. The partially digested feed bolus is later regurgitated, or brought back to the mouth to be chewed and swallowed again (Bakke et al., 2010).

Eructated gases, namely methane and carbon dioxide, are produced through microbial fermentation in the rumen. Peristalsis refers to the rhythmic contractions and relaxations of muscles that propel food through the esophagus, stomach, and intestines (Bakke et al., 2010).

Ruminants are animals that belong to the order Artiodactyl and have a specialized digestive system that allows them to extract nutrients from plant material through a process

Ruminants are a group of herbivorous mammals characterized by their hooves. This assemblage comprises giraffes, camels, llamas, deer, sheep, and goats. The stomach of a ruminant is exceptionally large. It fills 25% of their abdominal cavity. Their stomach is composed of four chambers, and food passes through them in a sequential manner (Figure 8.3):

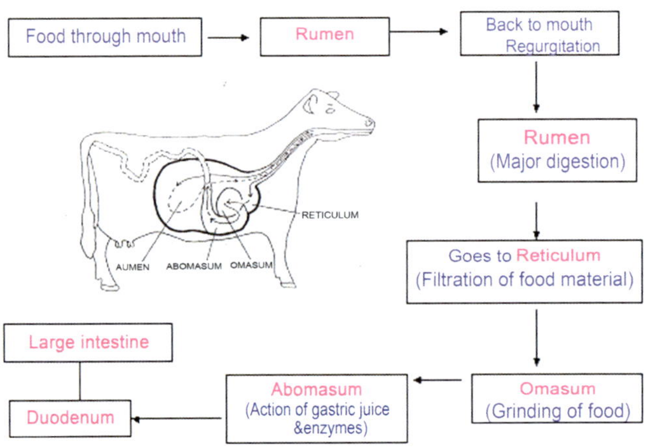

Figure 8.3. *The general features of a ruminant GI tract.*

Source: Muley (2020).

1. **Rumen:** Among the four chambers, the rumen is the most voluminous. Less dense foods, such as grass, are digested first in this process. Food is stored for future consumption. The rumen contains microorganisms capable of enzymatically degrading the cellulose and hemicellulose present in the plant matter that constitutes the majority of a ruminant's diet. The rumen provides optimal conditions for the proliferation

of microorganisms, including favorable pH and temperature levels. According to Bakke, Glover et al. (2010), animals raised for ruminant purposes exhibit infrequent chewing of their food. Partially ingested food can be regurgitated from the rumen and rechewed in the mouth. Regurgitation, also known as chewing cud, is the term used to describe this process. Not all animals engage in regurgitation, including ruminants. In addition, mothers engage in regurgitation to provide nourishment for their offspring (Bakke et al., 2010).

2. **Reticulum:** More compact materials twigs, go directly into the reticulum. Regurgitated food also enters the reticulum. The smaller food particles are propelled forward by the reticulum, while the larger items are retained for subsequent digestion.
3. **Omasum:** The compartment referred to as the digestive system receives water from the partially digested food. The interior of the omasum is characterized by numerous folds that bear a striking resemblance to the pages of a book. Similar to the pili found in the human digestive tract, these folds increase the surface area.
4. **Abomasum:** This is the rumen, which is the primary stomach of ruminant animals. Like non-ruminants, this very acidic chamber contains digestive enzymes. Like other mammals, food passes from the abomasum into the small intestine.

8.3.2.1. Microorganisms Found in the Rumen

Rumen microorganisms, comprising bacteria, fungi, protozoa, archaea, and viruses, aid in the process of digestion within the rumen.

The normal functioning of rumen microbes necessitates specific requirements, including an anaerobic environment devoid of oxygen, a pH range of 5.5–6.5 (slightly acidic), a temperature of 38–42°C, regular feed intake to facilitate fermentation, the elimination of waste products to prevent harm to the microbes due to acidity, and the maintenance of osmotic conditions. The growth requires volatile fatty acids, sufficient minerals (sodium, copper, cobalt, and phosphorus), nitrogen, and easily digestible carbohydrates for energy. The rumen bacteria are accountable for the processes of cellulose digestion, protein degradation, vitamin synthesis, and amino acid production (Bakke et al., 2010).

8.3.2.2. Food Undergoing Absorption into the Bloodstream Within the Rumen

Diffusion refers to the passive movement of small molecules, such as volatile fatty acids, ammonia, carbon dioxide, and methane, through the gut wall and into the circulation, driven by a concentration gradient. No energy is necessary. Glucose and specific amino acids are taken in through the intestinal lining via active transport, which operates in opposition to the concentration gradient. Adenosine triphosphate, also known as ATP, serves as the primary energy currency in living organisms (Bakke et al., 2010).

8.3.3. Non-Ruminant Digestion

Monogastric animals, such as pigs and poultry, undergo microbial fermentation in their large intestine as a result of their hindgut fermentation. It is located downstream from the small intestine and stomach. These animals do not possess the bacteria and enzymes required for the decomposition of cellulose. It is necessary to provide concentrated rations or other easily digestible meals to them (Bakke et al., 2010). Figure 8.4 depicts the non-ruminant digestive tract.

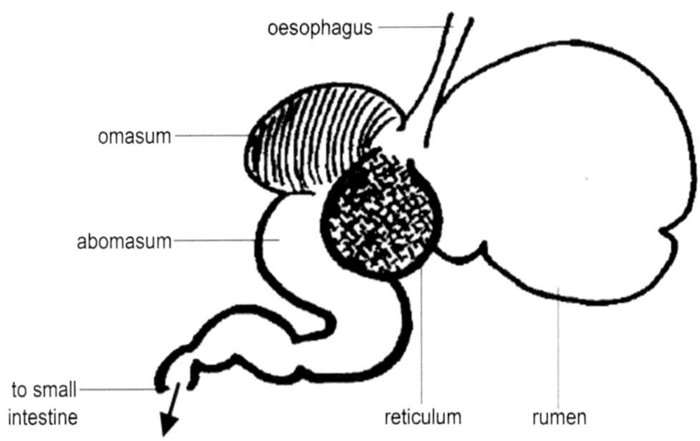

Figure 8.4. *Non-ruminant digestive tract.*

Source: Mbiza (2020).

The process of digestion involves the breakdown of intricate food molecules into smaller, more easily digestible molecules. Digestion takes place through two processes:

- Mechanically, where food is ground and chewed by the teeth, resulting in the formation of pulp.
- Chemically, by utilizing enzymes in the process of digestion.

The process of food digestion starts in the mouth, where it is initially mechanically and then chemically broken down by the salivary enzyme amylase.

Food is introduced into the stomach through the cardiac sphincter and undergoes digestion by enzymes and hydrochloric acid in gastric juice in the following manners:

- Pepsin converts proteins into peptides.
- Rennin converts soluble caseinogen into insoluble casein.
- Lipase produces glycerol and fatty acids from fat.

Chyme, which is food that has been partially digested, moves through the pyloric sphincter and enters the small intestine. In the small intestine, it is further broken down and almost completely absorbed with the assistance of pancreatic juice, intestinal juice secreted by glands in the intestinal wall, bile produced by the liver and stored in the gall bladder, and pancreatic juice (Bakke et al., 2010).

The ileo-cecal valve facilitates the passage of undigested food and waste materials into the large intestine, from where they are eliminated through the rectum and anus (Bakke et al., 2010).

The liver's exocrine gland produces alkaline bile, while the pancreas secretes pancreatic juice, exocrine pancreatic enzymes, and an endocrine component consisting of the Langerhans islands. These organs are classified as accessory in the context of digestion.

- **Gastrointestinal Glands:**
  - The Lieberkühn glands secrete succus entericus, which is a slightly alkaline juice found in the duodenum.
  - The duodenal glands and Brunner's glands secrete an alkaline fluid that contains a significant amount of mucus.

8.4. METABOLISM OF ESSENTIAL NUTRIENTS

The metabolism of essential nutrients is vital for animals to acquire energy for various life functions, including cellular metabolism, development, reproduction, and physical activity. Energy can be derived from the breakdown of proteins, fats, and carbohydrates present in food through a process called catabolism. Various forms of energy, such as radiant, chemical, mechanical, thermal, and electrical energy, have the ability to carry out mechanical, electrical, and chemical functions (Soetan et al., 2010).

Energy metabolism encompasses the breakdown and oxidation of proteins, fats, and carbohydrates within an organism's body, along with the subsequent release and utilization of the liberated energy to support life. Adenosine triphosphate (ATP) is the main energy source for all metabolic processes in animals. It is produced when dietary ingredients are broken down and oxidized (Soetan et al., 2010). ATP supplies energy for a variety of processes, including muscle contraction.

Animals primarily absorb carbohydrates in their digestive tract. Gluconeogenesis is the process by which non-sugar molecules, such as glycerol, lactic acid, propionic acid, and glucose-producing amino acids, are transformed into glucose. This metabolic process takes place in the liver and kidneys. Non-ruminant animals mainly acquire sugar from starch, which is transformed into glucose via enzyme hydrolysis in the digestive system and absorbed through the small intestine. Ruminants, however, acquire sugar from cellulose, which is metabolized by rumen microbes to generate low-quality fatty acids, such as butyric, propionic, and acetic acids. Propionic acid is the main substance that is used to create glucose (Soetan et al., 2010). Glucose is the primary energy source in animals and is broken down through different pathways, such as the pentose phosphate pathway, glycolysis pathway, and aerobic oxidation pathway.

The aerobic oxidation pathway is the process through which glucose is fully oxidized in the absence of oxygen to generate carbon dioxide and water. The pathway described is the predominant mechanism by which sugar is broken down, supplying energy to the majority of cells. The process entails the transformation of glucose into pyruvate in the cytoplasm, which is then converted into acetyl coenzyme A in the mitochondria via the pyruvate dehydrogenase complex. Acetyl-CoA is subsequently metabolized into carbon dioxide and water via the tricarboxylic acid cycle oxidation, resulting in the production of NADH, $FADH_2$, and ATP (Soetan et al., 2010).

Glycolysis is an alternative pathway for breaking down glucose in the absence of oxygen. The process entails the conversion of glucose into lactic acid and the liberation of energy. The pathway described can rapidly provide the necessary energy for physiological processes during exercise or hypoxia. Glycolysis takes place in the cytoplasm and involves two stages: the decomposition of glucose into pyruvate and the transformation of pyruvic acid into lactic acid through the action of lactate dehydrogenase. The crucial enzymes participating in glycolysis are pyruvate kinase, phosphofructokinase, and hexokinase (Soetan et al., 2010).

The pentose phosphate pathway plays a crucial role in supplying a hydrogen donor for biosynthetic reactions that are essential for synthesizing fat, cholesterol, and other compounds. This pathway exhibits higher levels of activity in adipose tissue, where there is an active process of lipid production.

8.5. ANIMAL NUTRIENT REQUIREMENTS

Nutrient requirements can be classified into four primary categories: growth, reproduction, lactation, and maintenance. The quantities of calories, protein, minerals, and vitamins needed are calculated based on these constituents. Producers can adjust to changes such as a prolonged period of cold weather, moving to a hilly pasture, or the final trimester of pregnancy by understanding the different factors that affect demand (Nolan, 2007).

8.5.1. Maintenance

- The maintenance component of an animal's diet encompasses all the essential nutrients required for respiration, locomotion, digestion, thermoregulation, wound healing, and weight maintenance.
- The main factors that affect maintenance needs are weight, age, breed, physiological condition, level of physical activity, and environment.
- The maintenance requirements of an animal are directly proportional to its size, especially in terms of energy and protein.
- Breeds possessing exceptionally large muscles will necessitate greater maintenance compared to breeds with smaller muscles.
- Pregnancy and lactation increase basal metabolism, necessitating distinct maintenance strategies. Breeds with high milk production exhibit elevated maintenance requirements.
- Severe cold, heat, rain, or mud, along with heightened activity and challenging terrain, will elevate the energy needs for maintenance. The only authentic livestock animal that requires maintenance is a fully grown castrated male horse that does not participate in any form of physical exercise (Nolan, 2007).

8.5.2. Lactation

- Heavy lactation necessitates a greater amount of nutrients when compared to other states of production.
- The nutritional requirements for lactation are determined by the amount and composition of milk produced during peak lactation.
- According to Nolan (2007), when milk production increases and the fat and protein content in milk increases, the animals will have higher dietary requirements.

8.5.3. Growth

- The rate of growth is highest during the initial stages of life, but gradually declines until adolescence. After reaching maturity, the rate of growth slows down even further until the individual reaches its full size.
- The requirements for growth are influenced by the actual weight, average daily gain (growth rate), weight at maturity, and composition of gain.

- In essence, the composition of gain refers to whether an animal is accumulating more adipose tissue or muscle mass.
- Young animals have higher protein requirements because they are building more muscle than fat (Nolan, 2007).
- As an animal matures, it accumulates more fat and requires a greater amount of energy.
- Younger animals have higher nutritional requirements per unit body weight.
- According to Nolan (2007), the highest efficiency is achieved when growth rates are extremely high.
- Following a period of energy restriction, young animals experience compensatory growth, resulting in a faster-than-normal weight gain.
- Growth, in this context, pertains to the necessity of adult animals to gain weight in order to enhance their body condition score (Nolan, 2007).

8.5.4. Reproduction

- The requirements for reproduction are modified based on the expected birth weight and stage of gestation.
- The development of the fetus and maternal tissue is essential.
- Pregnancy typically does not impact nutritional requirements until the final trimester, when the fetus experiences rapid growth.
- Poor nutrition prior to breeding can result in reduced fertility or the inability to maintain fertility.
- Underfeeding during growth can cause a delay in sexual maturity.
- Fetal tissues are given priority over maternal tissues when it comes to the allocation of nutrients.
- The body can deplete its reserves.

8.6. DISORDERS OF NUTRIENT METABOLISM IN ANIMALS

Understanding the impact of nutrient metabolism disorders in animals, especially in cattle, is crucial for maintaining their overall health and productivity. These disorders can arise when animals do not have enough nutrients, such as energy, calcium, and magnesium, to meet their metabolic needs. In high-demand environments such as the dairy sector, it is not uncommon to encounter metabolic disorders that can have serious consequences, potentially leading to acute and fatal deficits (Kaneko et al., 2008).

There are several metabolic disorders that can have an impact on cattle. These include ketosis, hypomagnesemia, milk fever, and fat cow syndrome. During the post-calving period, ketosis usually manifests after a few days to a few weeks. It is marked by a decline in milk production and appetite, along with symptoms such as constipation and weight loss. The treatment for ketosis typically involves the administration of intravenous glucose and oral propylene glycol (Kaneko et al., 2008).

Milk fever typically manifests around the time of calving and is marked by a decrease in appetite, elevated body temperature, impaired coordination, and, if not addressed, can progress to a state of reduced consciousness (Kaneko et al., 2008). The treatment for milk fever requires the administration of calcium borogluconate through intravenous infusion.

An imbalance in feeding during the early stages of pregnancy can result in fat cow syndrome, which can cause nutritional stress until calving. Cows that are affected by this condition experience a decrease in strength, loss of appetite, and pass hard and small droppings, among other symptoms. The treatment for fat cow syndrome involves the administration of anabolic steroids, providing supportive care, administering intravenous glucose and fluids, and giving oral propylene glycol (Kaneko et al., 2008). Hypomagnesemia is frequently observed in cows that are producing a large amount of milk and are grazing on nutrient-rich pastures. It is also commonly seen in calves that are primarily fed milk. This information has been documented by Kaneko et al., in 2008. The acute phase of hypomagnesemia is marked by sudden staggering, convulsions, and even fatality. The treatment for hypomagnesemia follows a similar approach to milk fever, and it includes administering a subcutaneous injection of magnesium sulfate (Kaneko et al., 2008). It is of utmost importance to prevent these metabolic disorders in cattle. Making dietary adjustments during the late pregnancy and peak lactation phase is essential. It is crucial to seek veterinary and dietary guidance, particularly in cases where these disorders are widespread (Kaneko et al., 2008).

8.7. IMPORTANCE OF PROPER NUTRITION FOR THE HEALTH AND WELL-BEING OF LIVESTOCK ANIMALS

Understanding the importance of proper nutrition is essential for ensuring the health and well-being of livestock animals. Animals, much like us, need a well-rounded diet that encompasses all the necessary nutrients, fluids, minerals, and vitamins to promote their growth, development, and overall well-being. Proper nutrition is essential for animals to thrive, enabling them to grow, reproduce, and develop a strong immune system that can effectively combat diseases. The benefits of this extend beyond the well-being of the animals, contributing to increased profitability and sustainability in agriculture (Ross et al., 2020).

Ensuring that every ration provided to livestock animals meets their specific nutritional needs at every stage of development is of utmost importance. It is important to note that the nutritional requirements of various species, such as cattle and pigs, can differ greatly. In contrast, the diet of a nursing cow will vary from that of a calf (Ross et al., 2020). It is crucial to ensure that the nutritional requirements of every animal are adequately met, whether they are being fed commercial concentrates or on-farm food.

Enhancing livestock feed with valuable additives can prove to be a highly effective method for boosting the nutritional value of the feed. One way to address the high energy requirements of cattle for growth and lactation is by incorporating B_{12} supplements and salt-based additives that contain a variety of micro and macro minerals. Proper nutrition for livestock animals plays a crucial role in preventing malnutrition, deficiencies, and diseases. Livestock animals are prone to various diseases related to nutrition, primarily caused by deficiencies in minerals and vitamins or inadequate nutrition (Ross et al., 2020). Deficiencies and malnutrition can have a profound effect on animal growth, development, and productivity, and in extreme situations, can result in mortality.

One of the consequences of modern farming methods is the rise in diseases associated with inadequate nutrition. It is worth noting that animals that are given concentrates aimed at optimizing productivity may experience deficiencies in essential minerals and vitamins (Ross et al., 2020). It is crucial to ensure that livestock feed is adequately supplemented with the essential nutrients to prevent deficiencies and promote optimal health. Proper nutrition is of utmost importance when it comes to animal reproduction. Multiple studies have demonstrated a clear link between animal reproductive performance and nutrition, as evidenced by the research conducted by Ross et al. in 2020. The reproductive capabilities of animals are greatly influenced by the quality and quantity of their rations, as well as their feeding practices and the nutritional content of their meals. Mineral deficiencies have a significant impact on various aspects of cow health and reproduction. They can disrupt the balance of hormones involved in gestation and parturition, leading to complications such as placental retention and mastitis. Additionally, these deficiencies can also impair cow fertility. Poor nutrition can have serious consequences, such as stunted growth after birth, inadequate fetal development, and, in severe cases, high rates of calf mortality.

Chapter **9**

Oxidative Stress and Antioxidants in Animals

Contents

The subject of oxidative stress and antioxidants in animals explores the intricate equilibrium between beneficial antioxidant defenses and harmful reactive oxygen species (ROS). Oxidative stress arises when an organism produces more ROS than it can counteract with its antioxidant capacity. Typically, ROS are generated as secondary products of essential biological processes. Stress negatively impacts animal productivity and health, leading to the development of various diseases. Lykkesfeldt & Svendsen (2007) investigate methods to decrease oxidative stress, enhance farm animal productivity, and improve welfare. Comprehending the intricate interplay between oxidation and defense, which involves enzymatic antioxidants such as superoxide dismutase and non-enzymatic ones like vitamin C and carotenoids, is crucial for ensuring animal well-being. This chapter will provide a thorough examination of various topics including: an overview of oxidative stress in animals, the presence of reactive oxygen species in animal cells, the antioxidant defense systems in animals, the impact of oxidative stress on animal health, the role of antioxidants in animal health, nutritional strategies to combat oxidative stress, and future directions in antioxidant research.

9.1. AN OVERVIEW ON THE OXIDATIVE STRESS IN ANIMALS

Our planet originated from a condensed accumulation of cosmic debris around 4.5 billion years ago. The temperature, ocean, and atmosphere of this entity were in a primitive state, and it underwent frequent cataclysmic catastrophes (Lykkesfeldt & Svendsen, 2007). The initial life forms eventually discovered a more hospitable habitat in this inhospitable and dreadful environment, although it lacked oxygen. Oxygen initially appeared in the atmosphere through the process of solar radiation causing the breakdown of water (H_2O) into hydrogen (H_2) and oxygen (O_2). Subsequently, oxygen levels increased further due to photosynthetic processes. As a result, anaerobic organisms must adapt to utilize oxygen and manage any potential hazards associated with oxygen and its metabolic by-products (Lykkesfeldt & Svendsen, 2007).

Stress is the automatic response that occurs when an organism is unable to adapt to its environment. This text provides a concise overview of the adverse impacts of stress on growth, productivity, reproduction, and disease. The potential consequences could vary from discomfort to fatality. Oxidative stress arises when an organism surpasses its ability to generate reactive oxygen species (ROS) and reactive nitrogen species (RONS. Oxidative stress, as identified by Lykkesfeldt and Svendsen in 2007, has the potential to impact various medical disorders that can affect both animal production and the overall well-being of individuals. Farm animals commonly undergo oxidative stress as a result of inadequate stable confinement and overcrowding, subpar nutritional and health conditions, and handling practices that consider the specific production cycle of each species.

9.1.1. Induction of Oxidative Stress

The stressors can be classified into endogenous sources and exogenous sources. The production of reactive oxygen species (ROS) and reactive nitrogen species (RONS) is a vital mechanism in the damage caused by different diseases and infections caused by both pathogenic and non-pathogenic organisms. Figure 9.1 depicts thedevelopment of oxidative stress in farm animals.

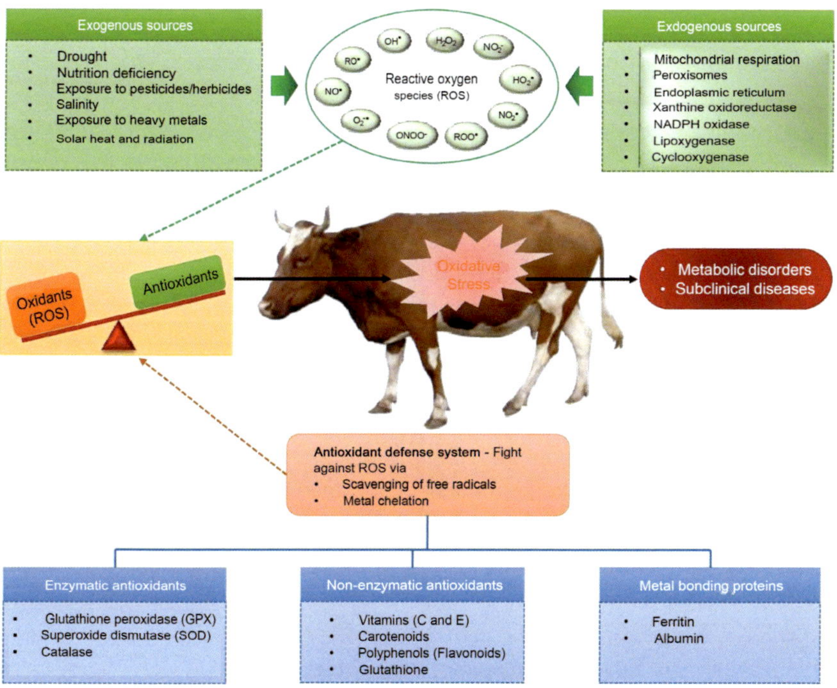

Figure 9.1.*Development of oxidative stress in farm animals.*

Source: Ponnampalam et al. (2022).

1. **Endogenous Sources:**The ROS/RONS is composed of both enzymatic and non-enzymatic components. These reactive species are generated through various biochemical processes and can function as intercellular signals in diverse biological processes and functions. The primary source of ATP is generated by the cell's

mitochondria through the process of respiration and mitochondrial oxidative phosphorylation. The main sources of reactive oxygen species/reactive nitrogen species (ROS/RONS) in mammals are the mitochondrial electron transport chain and the cytochrome P450 CYP-dependent ROS/RONS electron transport system. Mitochondrial NADPH oxidase (NOX) enzymes have the capability to produce reactive compounds, specifically superoxide anion (O_2^-), by utilizing cytoplasmic NADPH.

Cytochrome P450 (CYP) enzymes play a vital role in the activation of oxygen within the endoplasmic reticulum (Lykkesfeldt & Svendsen, 2007). They act as catalysts in NADPH-dependent reactions in the electron transport chain and generate $O_2 \bullet^-$. Cytochrome P450 is the main mechanism for detoxifying xenobiotics, which include drugs, pharmaceuticals, chemicals, alcohol, aromatic compounds, insecticides, and other industrial waste. Cytochrome P450 plays a crucial role in reducing various substances such as molecular oxygen, hydroperoxides, arene, N-oxides, azido and azodic compounds, halogen, nitrogenic compounds, hydroxylamines, and other xenobiotics. It also contains enzymes that facilitate the simultaneous reduction of molecular oxygen and the oxygenation of organic substrates. Some intermediates from the ROS/RONS reactions mentioned earlier lead to lipid peroxidation, which directly harms DNA and cellular membranes (Lykkesfeldt & Svendsen, 2007). Peroxidomes are the main sources of reactive oxygen species (ROS) and reactive nitrogen species (RONS).

Peroxisomes play a crucial role in the degradation of long fatty acid chains through beta-oxidation and the production of phospholipids, including plasminogen. These phospholipids are vital for the proper functioning of the heart, lungs, and brain in vertebrates. Moreover, in numerous eukaryotic cells, this particular subcellular organelle aids in the facilitation of specific enzymes involved in energy metabolism. In mammals, it also plays a role in the synthesis of cholesterol. Microglia cells are the main producers of reactive oxygen species/reactive oxygen and nitrogen species in the central nervous system. They release growth factors and anti-inflammatory cytokines to support the survival of neurons (Lykkesfeldt & Svendsen, 2007). Activated microglia cells produce nitric oxide, ROS/RONS, and other pro-inflammatory mediators. However, excessive activation of these cells can lead to various neurodegenerative disorders. As stated earlier, phagocytosis plays a crucial role in the production of reactive oxygen species/reactive oxygen and nitrogen species (ROS/RONS) and is closely associated with pathogenic infections (Lykkesfeldt & Svendsen, 2007).

An animal's oxidative balance can be disrupted by various organic, nutritional, and environmental factors, leading to a range of pathologies including neurological, cardiovascular, hepatic, renal, metabolic, autoimmune, and behavioral and reproductive disorders (Lykkesfeldt & Svendsen, 2007). Farm animals commonly undergo oxidative stress due to inadequate housing conditions, overcrowding in stables, subpar diet and health, and insufficient adherence to each species' production cycle (Lykkesfeldt & Svendsen, 2007).

2. **Exogenous Sources:**The organism can experience oxidative stress due to external factors that enhance the production of reactive oxygen species/reactive nitrogen species. Oxidative stress can be induced by various factors such as diseases, medications, radiation exposure, industrial chemicals, pollution, unfavorable living conditions, environmental factors, and pharmaceuticals (Lykkesfeldt & Svendsen, 2007). Research has demonstrated that reactive oxygen species/reactive nitrogen species (ROS/RONS), such as superoxide and hydroxyl radicals, play a crucial role in inducing apoptosis in infected cells. Pathogenic microorganisms have the ability to cause inflammation and destruction of infected cells and tissues. Experimental studies have shown that the purine degradation pathway, which is an important metabolic process for the production of reactive oxygen species (ROS) and reactive nitrogen species (RONS), plays a role in causing oxidative stress and inflammation in periodontal disease. The unfolded protein response (UPR) and bacterial infections are interconnected, indicating that pathogenic microorganisms induce oxidative stress in the endoplasmic reticulum. This suggests that this is the method by which nutrients are acquired from the host and prevents the process of internalization or permanent cell harm. Another example is idiopathic pulmonary fibrosis (IPF), a fatal lung disease with an unknown cause that is characterized by the gradual and chronic inflammation of the lung tissue, leading to the progressive deterioration of lung function. This condition is believed to be caused by reactive oxygen species/reactive oxygen and nitrogen species (ROS/RONS). Oxidative stress induces damage to different cells and tissues, as well as hypoxic conditions in the tissues (Lykkesfeldt & Svendsen, 2007).

9.2. REACTIVE OXYGEN SPECIES IN ANIMAL CELLS

Reactive oxygen species (ROS) are highly reactive substances formed from hydrogen peroxide, water, and diatomic oxygen (O_2) in the fields of biology and chemistry. Some common reactive oxygen species (ROS) include singlet oxygen, hydroxyl radical (OH), superoxide (O_2), and hydroperoxide (O_2H). ROS, or reactive oxygen species, are abundant due to their easy generation from oxygen (O_2). This makes them widespread in nature (Brieger et al., 2012). ROS, or reactive oxygen species, play a crucial role in a wide range of situations, encompassing both advantageous and detrimental effects. ROS function as molecular messengers that initiate and terminate biological processes. Oxygen (O_2) plays a crucial role in the redox reactions that occur in fuel cells. It acts as an intermediate in these reactions. The degradation of organic contaminants in the atmosphere through exposure to light is mainly influenced by reactive oxygen species (ROS). Most of the time, discussions about ROS primarily focus on biological aspects. These discussions explore topics such as the role of ROS in causing genetic changes that could be detrimental and their impact on the aging process (Brieger et al., 2012) (Figure 9.2).

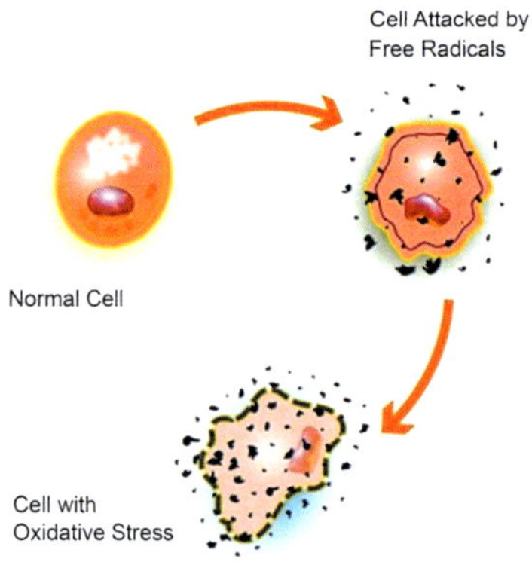

Figure 9.2.*High level of reactive oxygen species causes oxidative stress.*

Source: News-Medical (2020).

9.2.1. ROS Inventory

The definition of ROS can vary. Superoxide, singlet oxygen, and hydroxyl radicals are all considered to be sources. While hydrogen peroxide can be readily activated, it is not as reactive as these other species. Nitric oxide and peroxynitrite are examples of reactive oxygen-containing compounds (Brieger et al., 2012).

The Fenton reaction, involving hydrogen peroxide, ferrous compounds, and related reducing agents, generates the hydroxyl radical (HO). The chemical reaction between hydrogen peroxide (H_2O_2) and iron (II) (Fe (II)) produces iron (III) hydroxide (Fe (III)OH) and a hydroxyl radical (HO).

The hydroxyl radical rapidly and irreversibly interacts with all organic molecules during its short existence.

Oxygen (O_2) undergoes reduction to form superoxide ions (O_2^-). The mitochondria in the human body produce several grams on a daily basis.

The reaction is as follows: Oxygen (O_2) reacts with an electron (e^-) to form a negatively charged oxygen ion (O_2^-).

Superoxide dismutase are enzymes that facilitate the disproportionation of superoxide by competing with its formation in order to eliminate it (Brieger et al., 2012).

Hydrogen peroxide (H_2O_2) is a byproduct of respiration. Peroxynitrite (ONO_2^-) is formed through the combination of superoxide and nitric oxide. Singlet oxygen ($^1O_2$) is occasionally classified as a reactive oxygen species (ROS). Photosensitizers, like chlorophyll, have the ability to convert triplet oxygen ($^3O_2$) into singlet oxygen. Singlet oxygen exhibits a strong reactivity towards unsaturated organic compounds. Carotenoids, tocopherols, and plastoquinones present in chloroplasts effectively neutralize singlet oxygen and provide protection against its harmful effects. The presence of singlet oxygen can lead to the formation of oxidized products of β-carotene. These products can serve as second messengers, which have the ability to either protect against singlet oxygen-induced toxicity or trigger programmed cell death. Jasmonate levels are crucial in determining whether cells will adapt or die in response to increased levels of this reactive oxygen species.

9.2.2. Biological Function

Reactive oxygen species (ROS) are byproducts of regular oxygen metabolism in a biological context. They have specific functions in cell signaling and maintaining internal stability. Reactive oxygen species (ROS) are naturally occurring substances that play a vital role in the functioning of cells. They are found in normal cells at low and constant levels. Reactive oxygen species (ROS) play a role in plant metabolism, specifically in processes related to protecting against damage from light and increasing tolerance to different forms of stress. ROS, or reactive oxygen species, have the ability to induce permanent harm to DNA by oxidizing and altering certain cellular constituents, thereby impeding their ability to carry out their initial functions. This indicates that reactive oxygen species (ROS) have a dual function, where their harmful, protective, or signaling effects are determined by the equilibrium between ROS production and elimination at the specific time and place. This could have an impact on their behavior and ecology in terms of their ability to regulate body temperature, communicate visually, reproduce, and survive (Brieger et al.,

2012). Reactive oxygen species (ROS) can significantly increase in response to environmental stressors, such as heat or exposure to ultraviolet (UV) light. Cell structures can be severely damaged due to this. Oxidative stress is the term used to describe this phenomenon. ROS generation in plants is tightly controlled by stress factor responses. These factors, such as drought, salt, cold, pathogen defense, nutritional deficiencies, metal toxicity, and UV-B radiation, stimulate the production of ROS (Brieger et al., 2012). In addition, ionizing radiation from external sources generates reactive oxygen species (ROS), which have a permanent effect on tissue development in both plants and mammals.

9.2.3. Enzymes with Antioxidant Properties

1. **Superoxide Dismutase (SOD):** These enzymes facilitate the conversion of superoxide into oxygen and hydrogen peroxide. Consequently, in nearly all cells that are exposed to oxygen, they function as a vital antioxidant defense. Most chordates and mammals possess three distinct variants of superoxide dismutase. SOD2 is located within the mitochondria, SOD3 is situated outside of cells, and SOD1 is predominantly present in the cytoplasm. The remaining two are tetramers, consisting of four subunits each, while the first one is a dimer, composed of only two units. SOD2 contains a manganese ion in its reactive center, whereas SOD1 and SOD3 contain copper and zinc ions (Brieger et al., 2012).

The dismutation of superoxide, catalyzed by SOD, can be expressed through the following half-reactions:

$$M^{(n+1)+} + O_2^- \xrightarrow{SOD} M^{n+} + O_2$$

$$M^{n+} + O_2^- + 2H^+ \xrightarrow[SOD]{} M^{(n+1)+} + H_2O_2$$

M represents the elements Mn (with n = 2), Fe (with n = 2), Ni (with n = 2), and Cu (with n = The oxidation state of the metal cation varies between n and n + 1 in this process.

Catalase is located in the peroxisomes near the mitochondria and facilitates the conversion of hydrogen peroxide and water into oxygen and water. Glutathione peroxidase reduces the levels of hydrogen peroxide by transferring the energy of reactive peroxides to glutathione, which is a tripeptide containing sulfur. The transfer of reactive electrons from peroxide to glutathione is facilitated by the sulfur present in these enzymes, which acts as the reactive center. Peroxiredoxins located in the mitochondria, cytosol, and nucleus are also responsible for the decomposition of hydrogen peroxide (H_2O_2) (Brieger et al., 2012).

2. Detrimental Consequences:There are many species that have extensively studied the impact of reactive oxygen species (ROS) on cellular metabolism. These functions encompass apoptosis, which is the process of programmed cell death, as well as beneficial outcomes such as the activation of ion transporters and the stimulation of host defense genes. This implies that they have control over cellular activity. Reactive oxygen species (ROS) are specifically secreted by platelets that are involved in maintaining the balance of blood and promoting the healing of wounds. The purpose of releasing ROS is to attract additional platelets to the location of the injury. Leukocytes are recruited to establish a connection with the adaptive immune system (Brieger et al., 2012).

Reactive oxygen species (ROS) are associated with various inflammatory reactions, including cardiovascular disease, by affecting cellular activity (Brieger et al., 2012). Additionally, they could be associated with congenital deafness in both humans and animals, ototoxicity resulting from medications such as cisplatin, and hearing impairment caused by cochlear damage due to loud noises. ROS have also been associated with the initiation of ischemic damage and apoptosis, which refers to programmed cell death. Heart attacks and strokes are two distinct occurrences (Brieger et al., 2012).

Reactive oxygen species have the ability to harm DNA or RNA, oxidize polyunsaturated fatty acids in lipids (known as lipid peroxidation), oxidize amino acids in proteins, and lead to the oxidative inactivation of specific enzymes through oxidation co-factors. These actions have various harmful effects on the cell (Brieger et al., 2012).

9.3. ANTIOXIDANT DEFENSE SYSTEMS IN ANIMALS

For their levels to remain low but not zero, living cells must efficiently balance the processes of ROS/RNS production and inactivation/detoxification. For a long time, it has been understood that cells may repair the redox antioxidant/prooxidant balance and withstand mild oxidative stress by synthesizing more antioxidants (such as GSH, thioredoxin (Trx), antioxidant enzymes, CoQ, and others)). Because too much ROS can harm a variety of biological molecules, the antioxidant system network is in charge of preventing or reducing the harm. Nevertheless, the capacity of these adaptive mechanisms in living things is restricted. Oxidative stress results from the generation of more ROS/RNS than the antioxidant defense system can neutralize. This damages key biological molecules like proteins, DNA, and polyunsaturated fatty acids (PAFAs), and can have negative effects on health. As such, the antioxidant defense has a number of choices (Ighodaro & Akinloye, 2018) (Figure 9.3).

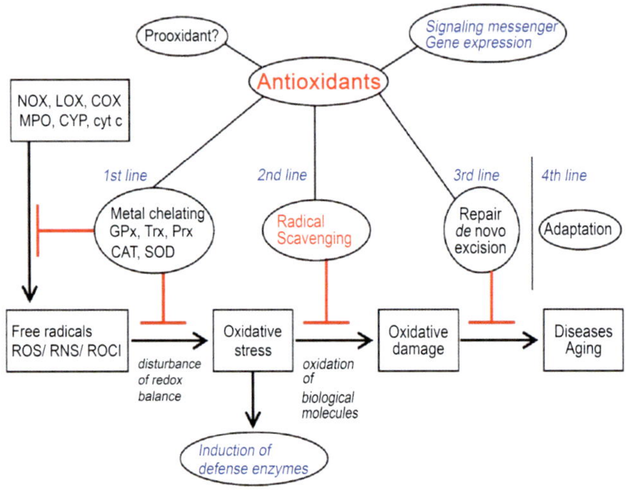

Figure 9.3.*Antioxidant defense mechanisms.*

Source: Sulaiman & Hussain (2011).

An antioxidant approach comes in various forms.

- First, efforts are being made to lower the production of free radicals by lowering the amount of oxygen available, lowering the activity of the enzymes that produce ROS and RNS (such as NADPH oxidase and xanthine oxidase), and keeping iron and copper bound to proteins to stop them from contributing to the formation of new radicals.
- Second, it is critical to preserve the integrity of mitochondria, which are the primary generator of free radicals in biological systems.
- Thirdly, an essential part of the antioxidant defense strategy is scavenging free radicals (e.g., vitamin E, vitamin C, GSH, coenzyme Q) and detoxifying/decomposing the free radicals and non-radical toxic compounds (SOD, GPx, catalase, etc.).
- Fourthly, the biological antioxidant effectiveness of vitamin E can be increased by the system of vitamin E recycling, which includes ascorbic acid, thioredoxin reductase (TrxR), vitamins B1 and B2, and vitamins B2.
- Fifth, the key components of the anti-stress approach are redox signaling, transcription factor (Nrf2) and vitamin E activation,

as well as additional manufacture of protective molecules with antioxidant and detoxifying properties.

- Sixth, the prevention of the build-up of damaged molecules and the maintenance of proteostasis are significantly aided by enzymatic systems that are in charge of damaged molecule repair (heat shock proteins, HSP; methionine sulfoxide reductase, Msr; DNA-repair enzymes, etc.) and removal (phospholipases, phospholipid hydroperoxide GPX (PH–GPx), proteasomes, etc.).
- Lastly, key components of the antioxidant defense network include apoptosis, autophagy, and other processes that deal with terminally injured cells to eliminate them and stop damages from spreading to other cells or tissues(Ighodaro & Akinloye, 2018).

An increasing amount of data indicates that the body's antioxidants function as a "team" to provide adaptive homeostasis (Ighodaro & Akinloye, 2018). When one team member assists another in working more effectively, there are cooperative exchanges. Antioxidant defense systems, which include both dietary and endogenous antioxidants (such as vitamin E, carotenoids, synthetic antioxidants, carnitine, silymarin, etc.) are demonstrated to be present in all cell compartments (such as the nucleus, cytoplasm, mitochondria, etc.) and to be expressed in a tissue-specific manner. Numerous antioxidant enzymes, such as SOD, GPx, and a few other selenoproteins, are induced by stress, and the level of stress affects both their expression and function.

9.3.1. The Concept of Oxidative Stress

An imbalance between oxidants and antioxidants, along with oxidative stress responses, was initially proposed by Sies in 1985 and subsequently revised (Ighodaro & Akinloye, 2018). The words "oxidative distress" and "oxidative eustress" were coined to describe severe and mild physiological oxidative stress, respectively (Figure 9.4.).

As a result of energy synthesis in mitochondria, reactive oxygen species (ROS) are normally produced and serve as a powerful "weapon" for phagocytes. To keep ROS levels low, the antioxidant defense network scavenges them and converts them into non-toxic chemicals, among other things. The interaction between specific targets and low ROS concentrations is essential for redox signaling, also known as oxidative eustress. These reactive oxygen species (ROS) work in tandem with several transcription factors and vitamin E to maintain homeostasis, adapt to stress, and protect health. Damaged proteins, DNA, polyunsaturated fatty acids (PUFAs), and other molecules are among the many untargeted targets affected by extensive reactive oxygen species (ROS) exposure due to compromised antioxidant defenses or excessive ROS production as a consequence of severe stress. Because of this, pathophysiology worsens, immunity is impaired, resistance to many diseases is reduced, and an animal's productivity and ability to reproduce are diminished.

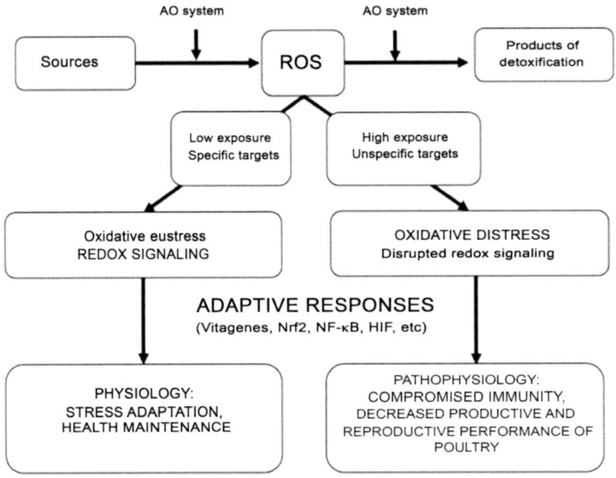

Figure 9.4. *Oxidative stress and adaptive responses.*

Source: Surai et al. (2019).

When faced with an oxidative stress, the body goes into stress mode to prevent ROS/RNS overproduction and set up ROS/RNS signaling in a way that promotes redox equilibrium. According to Ighodaro & Akinloye (2018), two important "master regulators" of the stress response were believed to be NF-κB/IkB and Nrf2/Keap1. A research paper on the subject was published in 2018. To be more precise, research has shown that both transcription factors are transported to the nucleus and bind to specific regions of DNA in response to stress, but they often have opposite effects. Ighodaro & Akinloye (2018) found that the stress response is generally an intricate process that is linked to other important biochemical and physiological pathways, such as the unfolded protein response, the hypoxia-induced response, and the heat shock response. Additionally, various processes such as autophagy, mitophagy, apoptosis, necroptosis, ferroptosis, and others have a substantial influence on homeostasis maintenance (Ighodaro & Akinloye, 2018).

9.4. THE IMPACT OF OXIDATIVE STRESS ON THE WELL-BEING OF ANIMALS

Oxidative stress occurs when there is an imbalance between the levels of antioxidants and free radicals in the body. During metabolic processes, the body generates free radicals, which are subsequently neutralized by antioxidants. Conversely, an excessive level of free radicals can impede the process of cell renewal and repair, leading to the earlier development of certain diseases and speeding up the aging process. Oxidative stress refers to the state of imbalance caused by an increase in oxidation sources and a decrease or impairment in antioxidant mechanisms. This definition asserts the notion that there exists a distinct type of mechanism, and any deviations from it could potentially affect homeostasis and potentially even trigger or worsen disease (Sordillo & Aitken, 2009).

Oxidative stress has an impact on chronic illnesses such as cancer, diabetes, neurological diseases, and cardiovascular disease. Extended exposure to elevated levels of pro-oxidant substances can lead to changes in the functioning of various enzymes and cellular structures, as well as defects in the DNA of the mitochondria, ultimately causing abnormal gene expression. In response, various endeavors have been undertaken to restore this "equilibrium" by administering antioxidants to both humans and animals, with the aim of safeguarding them against the detrimental impact of free radicals generated by bodily tissues. Oxidative stress, which can be either an epiphenomenon or a disease driver, negatively impacts the body's condition and heightens the requirement for antioxidants. Oxidative stress, which is the main cause or secondary effect of a disease, leads to the deterioration of bodily conditions. Antioxidant therapy can mitigate the occurrence of oxidative stress, which plays a vital role in regulating inflammation and cellular defense mechanisms (Sordillo & Aitken, 2009).

Oxidative stress refers to the disparity between the antioxidants and free radicals in the body. Molecules that contain oxygen and have an odd number of electrons are referred to as free radicals. Because of their disparate quantity, they are more prone to reacting with other molecules and initiating intricate chain reactions. Oxidation refers to these processes, which can have either beneficial or harmful effects. Oxidative stress, which occurs naturally, plays a vital role in numerous fundamental bodily processes. Long-term oxidative stress, caused by molecular and cellular damage, contributes to the onset of various acute and chronic disorders such as diabetes, cancer, heart disease, and neurological diseases (Sordillo & Aitken, 2009). Mitochondria are the primary organelles responsible for generating energy through oxidative phosphorylation to produce adenosine triphosphate (ATP).

Antioxidant enzymes effectively neutralize free radicals. However, imbalances can occur during critical periods, leading to an excessive production of free radicals that surpasses the capacity of antioxidants to neutralize them. Oxidative stress is a biological process that interferes with signal transduction, controls redox reactions, and leads to damage in biomolecules such as lipids, proteins, and DNA. Oxidative stress is characterized by an inequilibrium between the activities of pro-oxidants and antioxidants. Enzymatic and non-enzymatic systems, along with other substances that have oxidative potential, can effectively neutralize various reactive oxidative species. Oxidative stress has been linked to a variety of harmful biological effects, such as the resolution of inflammation and adaptive responses. However, instead of dealing with respiratory oxidative stress as a whole, implementing strategies that consider both individual and environmental risk factors could significantly enhance the effectiveness of antioxidants in therapy (Sordillo & Aitken, 2009).

Oxidative stress *in vivo* leads to the generation of reactive oxygen species (ROS) and the activation of antioxidant defense mechanisms. Reactive oxygen species are inevitably generated as a result of normal aerobic metabolic processes in the mitochondrion. Mitochondrial oxidative phosphorylation supplies energy to spermatozoa, oocytes, and embryos by means of reactive oxygen species. Oxidative stress arises from an overabundance of reactive oxygen species (ROS), which impairs the cells' capacity to act as antioxidants. Oxidative stress is also associated with the adverse consequences of livestock infertility. ROS exhibit a positive correlation with all forms of stress encountered by animals. There is a correlation between an increase in reactive oxygen species (ROS) in the male reproductive tract and stress. Antioxidants and oxidative stress have become crucial subjects in numerous scientific discussions regarding issues related to the study of bioscience. Free radicals and oxidative stress occur due to a disruption in redox signaling and regulation, leading to the impairment of macromolecules such as proteins, lipids, and DNA. This interference with normal metabolism and physiology ultimately leads to necrosis or the loss of cell function (Sordillo & Aitken, 2009).

The primary constraints on animal productivity in tropical and subtropical regions are elevated ambient temperature and humidity. Oxidative stress is quantified by measuring the levels of oxidants and antioxidants, which indicate an imbalance between the two. Heat stress often leads to oxidative stress in animals in tropical regions, resulting in the production of Reactive Oxygen Species (ROS). The increased production of reactive oxygen species (ROS) and a decline in antioxidant defense mechanisms are responsible for the emergence of various diseases, such as asthma and other respiratory issues. Oxygen-centered free radicals and their metabolites, commonly referred to as "reactive oxygen metabolites" (ROMs), are the predominant free radicals found in biological systems). Regular metabolic processes such as protein phosphorylation, transcription factor activation, cell differentiation, apoptosis, oocyte maturation, steroidogenesis, cellular immunity, and microbial cellular defense contribute to the continuous formation of reactive oxygen metabolites

(ROM). In small amounts, it is essential for various physiological processes (Sordillo & Aitken, 2009).

9.5. THE ROLE OF ANTIOXIDANTS IN ANIMAL HEALTH

The utilization of antioxidants in animal nutrition has increased due to the current inclination towards formulating diets abundant in polyunsaturated fatty acids (PUFAs), which are highly susceptible to lipid peroxidation. An antioxidant is a substance that can effectively inhibit or prevent the oxidation of food when it is present in the food at a concentration lower than the oxidizable matrix. Antioxidants are incorporated as additives in the production, processing, and preservation of animal feeds, premixes, and fats to prolong their shelf life by inhibiting lipid peroxidation and oxidative rancidity (Spears & Weiss, 2008).

9.5.1. The Technological Function of Antioxidants in Animal Nutrition

Oxidation reduces the energy provided by the fats and oils in the feed that is consumed by animals, and it also greatly decreases the nutritional value of the feed (Spears & Weiss, 2008). Similarly, feed that has undergone oxidation contains a significant amount of free radicals and peroxides, which can pose a risk to animals. Antioxidants play a crucial role in protecting food items such as feed, flours, premixes, and animal fats from oxidative damage, thereby maintaining their sensory properties and extending their shelf life. This serves to inhibit the occurrence of rancidity and discoloration. Antioxidants are employed in animal nutrition to help preserve the sensory characteristics of the feed and prevent the degradation of essential nutrients such as amino acids. Businesses constantly seek innovative antioxidants that do not affect the sensory qualities of products, are safe for human health, and are economically viable. Furthermore, it is necessary for them to possess liposolubility, exhibit efficacy at low temperatures, and demonstrate resistance to feed treatments (Spears & Weiss, 2008).

9.5.2. Biological Function of Antioxidants in Animal Nutrition

Antioxidants have the important role of inhibiting the oxidation of chemicals that can alter physiological processes and are essential in protecting against diseases caused by oxidative stress (Spears & Weiss, 2008). Oxidative stress can be defined as the condition that occurs when there is an imbalance between the body's natural defense mechanisms against free radicals and the production of free radicals within the body. Consuming antioxidants from external sources is essential when there is an excessive production of free radicals to prevent potential damage to cells. The balance between oxidation and antioxidation is primarily influenced by exogenous antioxidants. According to Spears & Weiss (2008), maintaining normal levels of antioxidants generally has beneficial effects.

Animals possess a biological system of antioxidants that aids in combating free radicals, which are continuously generated by their own metabolic processes (Spears & Weiss, 2008). The innate antioxidant defense system offers protection, albeit with certain limitations. The presence of certain elements that can cause an excessive production of free radicals and/or hinder the ability of the biological system to function as an antioxidant, leading to oxidative stress, further endangers this threshold. These factors include the consumption of mycotoxins, heavy metals, fungicides, pesticides, feed rich in polyunsaturated fatty acids, insufficient nutrition, and pathogenic diseases.

Antioxidants are essential for optimizing animal yields in livestock production. This is because they not only increase the economic output, but also enhance the quality of the products (such as milk, meat, eggs, etc.) and guarantee customer safety. Animals can receive antioxidants in a cost-effective, practical, and effective way by incorporating them into their diet.

Furthermore, certain antioxidants, such as tocopherols, not only provide technological protection but also offer biological advantages due to the presence of alpha tocopherol (Vitamin E) in their composition. The combination of alpha, beta, gamma, and delta isomers results in the formation of tocopherols. The relative proportion of each isomer varies depending on the source of the tocopherols.

Research has shown that incorporating different levels of d-alpha tocopherol into the diet of ruminants has positive effects in reducing the incidence of diseases. Dairy cows have been administered antioxidant vitamin supplements in order to reduce the likelihood of mastitis infection.

Vitamin E prevents the rapid oxidation of oxymyoglobin in meta myoglobin, a protein that contributes to the sensory qualities of meat. This preservation of color in meat after slaughter is an enhancement compared to meat from animals that have not been given vitamin E supplements.

Although antioxidants have been commonly used in food processing, incorporating antioxidant supplements into animal nutrition can be a more economical and effective method to improve the oxidative stability, sensory characteristics, and antioxidant nutritional value of animal products.

Feeding animals with antioxidant-enriched food delays the occurrence of discoloration on the surface of muscles and reduces the production of meta myoglobin in meat intended for human consumption (Spears & Weiss, 2008).

Prior to receiving approval for use, each antioxidant utilized in the European Union undergoes a meticulous testing procedure in laboratories to guarantee that it does not have any adverse effects on animal health. After receiving approval, the usage of these items is usually limited to specific meals and quantities (Spears & Weiss, 2008). Furthermore, in accordance with EU regulations, it is mandatory for all food additives, including antioxidants, to be clearly specified on product labels.

9.6. STRATEGIES FOR ADDRESSING OXIDATIVE STRESS THROUGH NUTRITION

Animals reared in intensive care settings often encounter oxidative stress, leading to increased feeding costs and heightened susceptibility to health issues. Therefore, it is crucial to develop effective strategies for reducing oxidative stress.

During oxidative stress, the body's nutrient metabolism undergoes a change in direction. To alleviate this stress, the use of dietary supplements or non-nutritional additives can be beneficial (Sies et al., 2005). The commonly used supplements can be classified into the following categories: 1. Amino acids such as cysteine, arginine, and tryptophan; 2. Vitamins including A, C, and E; 3. Mineral elements such as copper, manganese, zinc, selenium, and others; 4. Natural compounds like resveratrol and curcumin (Sies et al., 2005).

9.6.1. Functional Amino Acids

Functional amino acids, such as cysteine, have been found to significantly decrease intestinal damage and inflammation caused by lipopolysaccharide (LPS). Additionally, cysteine supplementation has been shown to increase the levels of GSH (glutathione) in the gut, as well as enhance the activity of antioxidant enzymes such as SOD (superoxide dismutase), GSH-Px (glutathione peroxidase), and catalase. These findings were observed in a piglet model. Piglets exhibiting developmental delays demonstrated significantly improved growth and development when supplemented with 1.2 g/kg of a cysteine precursor known as acetylcysteine. This intervention increased the level of GSH and increased the expression of genes associated with the production of GSH in the live. The addition of acetylcysteine to the diet has been found to have significant anti-inflammatory effects on oxidative stress and inflammation caused by lipopolysaccharide in piglets (Sies et al., 2005).

Arginine possesses several indirect mechanisms to regulate the body's oxidation system. Arginine significantly increased the activity of GSH-Px and SOD in the piglet model of oxidative stress and decreased the oxidative damage caused by diquat. Experiments conducted on pig finishing showed that the addition of arginine resulted in an improvement in the muscle and meat quality indexes of pork, specifically in terms of antioxidant capacity. A recent study conducted on IPEC-J2 cells demonstrated that L-arginine reduced the oxidative stress and inflammatory response induced by lipopolysaccharide through the activation of arginase-1 signaling.

9.6.2. Vitamin-Based Antioxidant Supplementation

Vitamin E, the primary fat-soluble antioxidant, aids the body by modulating cytokine expression and reducing oxidative stress. Studies have demonstrated that Vitamin E has the ability to mend DNA damage in sperm, enhance the success rates of *in vitro* fertilization, increase the number and viability of sperm in animals, protect against sperm damage and harm to embryos caused by PVC, and promote the development of embryos. In a separate study, piglets that were given peroxidized soybean oil exhibited symptoms of oxidative stress and experienced growth impairment. However, when vitamin E was added as a supplement, it significantly increased the levels of vitamin in the blood serum compared to the control group and the group treated with polyphenols (1.98, 1.25, and 1.26 mg/kg) ($p < 0.001$). The study found that the antioxidant capacity of piglets was enhanced by both vitamin E and polyphenols. However, there were no significant changes in growth performance or oxidative state compared to the control group.

Vitamin C acts as an antioxidant through the vitamin E-dependent pathway. It helps to remove reactive oxygen species (ROS), which prevents damage to lipids and proteins and protects cells from harm (Sies et al., 2005). Supplementing with Vitamin C, regardless of the dosage (low or high), significantly improved fish growth and feed efficiency, as well as increased the weight of the animals. A 2020 review of the impact of vitamin C on metabolic syndrome found that certain beneficial effects of the vitamin were influenced by its anti-inflammatory and antioxidant properties. Vitamin C treatment significantly increased the levels of SOD and GPx, while decreasing the levels of MDA, when compared to the control levels ($p < 0.05$). The study found that the combination of fipronil and pirimiphos-methyl insecticides had a reduced toxic effect on the nervous system, decreased the harmful effects of oxidative stress, and improved the movement ability of young zebrafish. These effects may be attributed to changes in the expression of genes related to the hypothalamic-pituitary-gonadal axis and increased cortisol levels (Sies et al., 2005).

9.6.3. The Mineral Elements

Copper is essential for the body to counteract oxidative stress, as it plays a crucial role in the activity of the copper-scavenging enzyme Cu/Zn SOD, which helps neutralize oxidative stress. The presence of cholesterol in an animal's diet can worsen oxidative damage due to a shortage of copper. The copper shortage has the potential to impact animal microbiomes. The addition of copper nanoparticles to the meals of weaned piglets resulted in an enhancement of their antioxidant capacity. This indicates that administering copper nanoparticles at a dosage of 100 mg/kg body weight to weaned piglets could potentially serve as a feasible alternative (Sies et al., 2005).

Selenium primarily demonstrates its antioxidant properties by participating in the activity of GSH-Px, a crucial antioxidant enzyme that protects mammalian cell membranes from oxidative harm. Oxidative stress occurs as a result of a deficiency in selenium, which is regulated by selenoproteins. Consequently, this triggers inflammatory reactions *via* the NF-κB, HIF-1α, and NLRP3 pathways, ultimately leading to harm to the organs. The study conducted by Sies et al. (2005) examined the inclusion of various selenium sources, such as selenium chelated soy protein, sodium selenite, and selenium-treated yeast. The nursery piglets were categorized into five groups to investigate the oxidative stress induced by diquat. The first group, known as the negative control (NC), received an intraperitoneal injection of sterile saline after being fed a basic diet. The first group is the positive control (PC), which is provided with a standard diet and administered an injection of diquat solution. The remaining four options include soy protein-chelated selenium (SC), a basal diet with 0.3 mg/kg of selenium from sodium selenite (SS), and selenium yeast (SY). After conducting the experiment, it was determined that adding a selenium-enriched yeast source with a concentration of 0.3 mg/kg selenium had the most positive effects. This is because the organic selenium significantly increased the activity of natural antioxidants in all areas, as compared to the control group ($p < 0.05$). According to Sies et al. (2005), SY achieved the highest level of success in this aspect. A new variant of selenium nanoparticle, known as biogenic nanoselenium (BNS) particles, has been found to provide better protection to the mouse intestinal barrier and maintain intestinal redox homeostasis compared to Se-Met and Nano-Se in a mouse model of intestinal oxidative stress (Sies et al., 2005).

9.6.4. The Natural Compounds

Curcumin, found in turmeric, possesses anti-inflammatory, antioxidant, antiproliferative, and antiangiogenic properties. Curcumin can scavenge free radicals and act as an antioxidant by activating the Nrf2 signaling pathway. Piglets with intrauterine growth retardation (IUGR) exhibited decreased levels of oxidative stress in their intestines and increased levels of antioxidant capacity when their basal diet was supplemented with 400 mg/kg of curcumin. In Cao's study, it was demonstrated that curcumin effectively reduces oxidative stress, enhances intestinal barrier function, and improves mitochondrial function by activating AMPK and promoting TFEB nuclear translocation, which leads to the induction of Parkin-dependent mitophagy (Sies et al., 2005).

9.7. PROSPECTS FOR ADVANCEMENTS IN ANTIOXIDANT RESEARCH

In recent years, the importance of nutrition has greatly grown, and antioxidants have emerged as crucial elements in the development of functional meals that are necessary for preserving animal health and preventing illness. Antioxidants can be synthesized in a laboratory or obtained from various natural sources. The state of equilibrium between oxidants and antioxidants is referred to as "oxidative balance," and it is crucial for overall health. In contrast, oxidative stress, resulting from an imbalance that favors oxidants, can lead to cellular damage and contribute to the development of pathological conditions. Recent cancer research has shown increased interest in the impact of dietary antioxidants (Firuzi et al., 2011).

Antioxidants in animal diets not only promote animal health and reduce the occurrence of disease, but they also have the capacity to decrease the overall reliance on antibiotics in livestock. Antioxidants help decrease the need for antimicrobials by bolstering the immune system and improving overall health. In addition, sustainable agricultural techniques advocate for the utilization of antioxidants as a means to combat antibiotic resistance and support public health and food safety efforts.

The bioavailability of antioxidant compounds can be influenced by various factors specific to each species, such as the differences in digestive systems between ruminants and monogastrics, the enzymatic portfolio, feed composition, and microbiota contributors (Firuzi et al., 2011). Over the past decade, numerous natural extracts and compounds have been introduced to the market as functional ingredients or additives. Nevertheless, due to the diverse range of antioxidants available and the varying dosages used, researchers have encountered contradictory findings in the literature. This highlights the need for further comprehensive research to thoroughly characterize each antioxidant and determine the most optimal inclusion in animal feed (Firuzi et al., 2011).

In addition, Laviano conducted an assessment of the effects of adding hydroxytyrosol to the colostrum and milk of Iberian sows during the late stages of pregnancy. Based on this study, the addition of hydroxytyrosol led to a reduction in the ability to convert fatty acids in 20-day milk, while increasing the levels of polyunsaturated fatty acids in colostrum. The alterations in the composition of milk lipids can potentially benefit the oxidative condition and gastrointestinal well-being of piglets during their early stages of life.

The addition of selenium nanoparticles and organic selenium supplementation at a dosage of 0.5 mg/kg resulted in enhanced physico-chemical quality of the breast, improved animal health, and increased antioxidant capacity of the liver and muscles in broilers). The study conducted by Firuzi et al., in 2011 found that nano-sized Se did not cause any harmful effects and also showed a preventive effect against mitochondrial damage in hepatocytes. Broiler chickens have been administered a dose of 600 mg/kg of Myricetin, a naturally occurring flavonoid. Research has demonstrated that this approach leads to a notable enhancement in the growth performance and antioxidant capacity of the meat. Additionally, it reduces the relative expression of interleukin, the severity of lesions caused by Eimeria spp., and oocyst shedding.

The collective findings of the published studies emphasized the importance of optimizing the advantages of natural antioxidants in animal nutrition. Scientists are conducting research to establish sustainable approaches for raising livestock by studying various bioactive substances and their mechanisms of action. An emerging area for enhancing animal health, welfare, and productivity involves incorporating natural antioxidants into animal feed, as part of our efforts to adopt a more ethical and comprehensive approach to livestock production (Firuzi et al., 2011).

Chapter **10**

The Impact of Environmental Factors on Animal Biochemistry

Contents

The relationship between animal biochemistry and environmental influences is highly complex, as it directly impacts the ability of animals to adapt to their surroundings and manage stress. Gaining knowledge about the connection between plant poisons, pollutants, and metabolic pathways can offer valuable insights into the well-being and survival of animals. The relationship between biology and the environment is clearly demonstrated in different situations, like fish swimming in polluted waters or desert lizards sunbathing.

10.1. THE INFLUENCE OF ENVIRONMENTAL FACTORS ON ANIMAL METABOLISM

Environmental factors that affect animal health are predisposing factors, such as the physiological condition of dairy cows. The neonatal period, post pubertal reproduction, and nursing are crucial stages in life. During the newborn period, the primary consequence of the environment is an increased susceptibility to disease, which is associated with a decreased level of immunoglobulin in the plasma of calves (Willmer et al., 2009). Cattle experience a decrease in sexual drive, fertility, and the ability of embryos to survive when exposed to heat stress. However, cold stress does not have any impact on their reproductive functions. Exposure to heat in the later stages of pregnancy reduces fetal growth and alters the endocrine system of the mother. Heat stress during late gestation has noticeable impacts on postpartum breastfeeding and reproduction.

Heat stress significantly decreases rumination and intake of roughage in lactating cattle. Decreasing the intake of roughage can result in a reduction in the production of volatile fatty acids and potentially alter the acetate to propionate ratio. Heat stress leads to a decrease in rumen pH (Willmer et al., 2009). Cattle experiencing heat stress exhibit reduced levels of sodium and potassium in their rumen fluid, leading to lower serum electrolyte contents. Lowered sodium and potassium levels are associated with increased potassium and sodium loss through the skin and urine, as well as decreased levels of plasma aldosterone and elevated levels of plasma prolactin. Extended exposure to high temperatures leads to decreased levels of glucocorticoids, growth hormone, and thyroxine in cattle. These reductions in hormone concentrations appear to be linked to a decrease in basal metabolism (Willmer et al., 2009).

Climate is a variable that can significantly affect the status of a disease, and it is expected to have a severe negative impact on the health of both humans and animals. Moreover, several studies have suggested that increasing the temperature can reduce mortality and improve the well-being and health of both cattle and people living in areas with severe winters (Willmer et al., 2009).

The direct or indirect effects of climate change on animal health may primarily result from alterations in air temperature, relative humidity, precipitation, and the occurrence and severity of extreme events (such as heat waves, severe droughts, extreme precipitation events, and coastal floods).

10.1.1. Disorders of Metabolism

Reproductive issues can also arise as a result of heat stress in animals. Research has shown that elevated temperatures can have a detrimental effect on the reproductive capabilities of animals, affecting both males and females (Willmer et al., 2009). Heat stress in males can result in a decline in sperm quality, which includes a decrease in sperm motility and viability. Additionally, testicular degeneration and a decline in testosterone production can occur. Heat stress has the potential to disturb the estrous cycle in females, resulting in heat periods that are either irregular or prolonged. Additionally, it can lead to a decline in fertility rates and a rise in embryonic mortality, as noted by Willmer et al., in 2009. It is important to note that heat stress can have a significant impact on the development of the fetus. This can result in lower birth weights and higher neonatal mortality rates.

Heat stress can have negative impacts on the immune system of animals. Under elevated temperatures, the immune response becomes weakened, rendering animals more vulnerable to infections and diseases (Willmer et al., 2009). Heat stress has the ability to inhibit the production of immune cells and decrease the activity of immune system components, such as antibodies and cytokines. It is important to note that there can be a higher susceptibility to respiratory and gastrointestinal infections, along with other diseases.

Aside from the physiological impacts, it is important to consider the potential behavioral and welfare consequences that animals may experience due to heat stress. When animals are subjected to heat stress, they may display changes in their behavior. These changes can include restlessness, reduced activity levels, and alterations in their feeding patterns. It is worth noting that animals may also actively search for shade or water sources in order to regulate their body temperature. Behavioral changes can potentially have detrimental effects on the well-being of animals. Such changes can lead to discomfort, stress, and decreased productivity (Willmer et al., 2009).

Implementing various management strategies can help mitigate the negative effects of heat stress. It is essential to ensure that animals have access to shade and water sources in order to facilitate their cooling and hydration. Installing ventilation systems in barns and housing facilities can greatly enhance air circulation and effectively minimize heat buildup. In addition, making changes to feeding practices can be beneficial in reducing metabolic issues.

10.2. THE IMPACT OF POLLUTION AND TOXINS ON ANIMAL HEALTH

Improper livestock production intensification as part of the development process can result in various negative consequences. These include overgrazing, decreased soil fertility, erosion, and desertification. It is particularly evident in marginal agricultural areas, where the majority of extensively managed ruminant farms are situated. Efforts to enhance animal health, such as immunization campaigns or programs to control parasites like ticks or tsetse, have been found to have a positive impact on the size and productivity of animal populations (Hu et al., 2017). Failure to implement proper land-use planning can exacerbate land degradation and place additional strain on animal populations.

In order to mitigate the risk of negative ecological impacts and take into account the diverse agricultural, topographical, and geographical factors at play, it is imperative to implement effective land-use planning and management. This not only enhances productivity and prevents the spread of animal diseases, but also safeguards against undesirable environmental consequences. In order to ensure effective land planning and utilization, it is imperative to adopt a multidisciplinary approach.

10.2.1. Pollution

In addition to this, the rearing of livestock also results in the production of various waste materials, such as feedlot dung, and an uptick in the utilization of veterinary products like insecticides. When considering the development of new animal housing, especially in industrial production systems, it is crucial to address the increasingly concerning problem of animal wastes, such as manure and liquid manure, polluting the environment, especially water supplies (Hu et al., 2017). It is crucial to take the necessary measures to ensure the proper management and disposal of waste generated by slaughterhouses.

When handled correctly, these byproducts can prove to be quite valuable. Prior to further processing and utilization, it is necessary to sterilize or treat any substance that is considered unacceptable or contaminated. Improper disposal of this type of waste can lead to an increase in the population of predatory animal species, both in the sea (like sharks) and on land (such as hyenas and rural dogs) (Hu et al., 2017). It is crucial to sterilize refuse from international air and sea travel to prevent the spread of animal diseases through infected animal products. Currently, there exist environmentally friendly methods for the application of acaricides (pour on) and insecticides (targets and traps for tsetse control). It is highly recommended to utilize these whenever possible, as they possess the capability to reduce the chances of environmental pollution. Two examples of cattle breeds that exhibit resistance to parasites are trypanotolerant cattle and tick-resistant breeds. These breeds, or their crosses, have the ability to withstand parasite species, thereby reducing the reliance on pesticides (Hu et al., 2017).

Shifting the ecological balance is a complex and intricate process that has far-reaching implications. It involves a delicate interplay of various factors and can have significant consequences for the environment. Understanding the intricacies of this phenomenon requires a deep dive into the intricate web of ecological relationships and the intricate mechanisms that govern them.

By examining the various components of the ecosystem and their interconnectedness, we can When a species' population decreases in a certain area, it can lead to unexpected consequences for other species that were not the intended target. These effects, in turn, have an impact on the overall environment. Occasionally, the implementation of disease-control strategies can lead to unforeseen environmental consequences.

The emergence of drug-resistant pathogen strains has made control more challenging due to the disproportionate and extensive use of parasiticides and antibiotics, such as acaricides and anthelmintics. In Mexico, the act of poisoning coyotes in order to control rabies had an unintended consequence – the jackrabbit population began to rapidly increase, posing a significant challenge for the agricultural sector. African game parks have the potential to be a source of various livestock infections, including trypanosomiasis and foot-and-mouth disease (Hu et al., 2017).

10.2.2. The Use of Veterinary Products

It is of utmost importance to educate all individuals responsible for the management and distribution of veterinary products regarding the potential environmental hazards associated with these medications. This is particularly significant considering the increasing utilization of these drugs for disease treatment, parasite control, and productivity enhancement (Hu et al., 2017). When it comes to veterinary products, it is important to carefully follow the instructions provided.

These instructions cover how to use the products, how to properly dispose of them, and the potential negative effects they may have. Additionally, these instructions also offer guidance on how to handle and neutralize pesticides in case of an accident. The wording utilized for this material ought to be suitable. It is worth noting that the information on the packaging may vary across different countries due to national laws related to registration (Hu et al., 2017).

It is crucial for animal health services and registration authorities to collaborate closely in order to establish and uphold a superior standard for packaging and usage instructions. It is important to ensure that instructions for products intended for inexperienced individuals, such as animal owners, are provided in a language that is suitable for them. According to Hu et al. (2017), it is recommended that manufacturers submit proposed labels for new products to registration authorities for approval in advance.

10.2.3. Using Insecticides

It is crucial to have strict regulations in place for pesticide imports, registration, distribution, and use. Additionally, it is important for users to undergo adequate training on proper handling and application techniques.

10.2.4. The Presence of Toxic Residues in Animal Products

It is important to note that the edible products of treated animals may contain residues of veterinary drugs even after they have been administered (Figure 10.1). There are three distinct categories that we need to consider when it comes to potential health risks associated with food residues. These categories include immunopathology, microbiology, and toxicology. One important point to consider is the subtherapeutic usage of antibacterial agents in feed.

Additional chemicals are utilized in animal husbandry. Various additives are employed to preserve the quality of the feed. These additives encompass colorants, disinfectants, pesticides, antioxidants, and antifungal agents. It is worth noting that there are also significant concerns regarding public health. It is important to consider strengthening control measures for pesticide, drug, and hormone residues in meat, milk, eggs, and other animal products, even in developing nations where this practice is not yet widespread. It is crucial to conduct informational campaigns, conferences, and training initiatives to address the residues found in animal products. These measures are necessary to prevent and mitigate these issues, as highlighted by Hu et al. (2017).

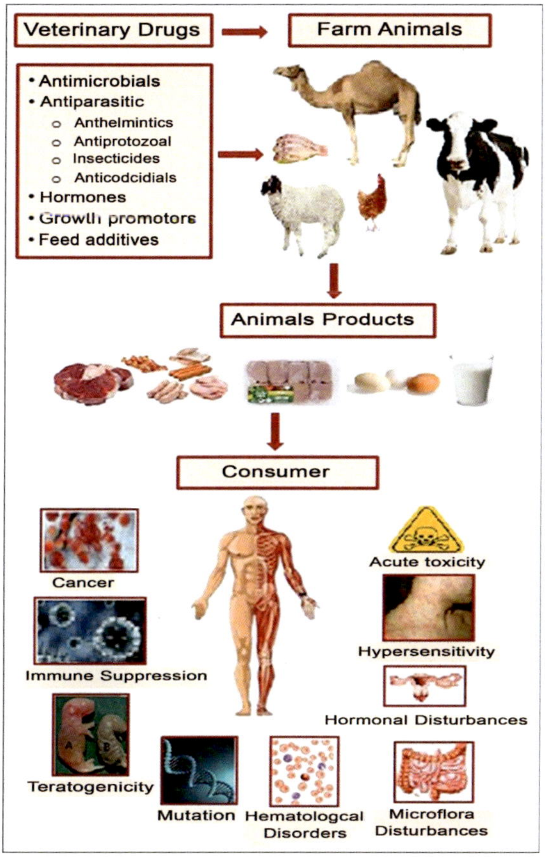

Figure 10.1.*Current perspective on veterinary drug and chemical residues in food of animal origin.*

Source: Atta et al. (2022).

10.3. CLIMATE CHANGE AND ANIMAL BIOCHEMISTRY

Anthropogenic climate change will continue to exert a dominant influence on ecosystem flux for the foreseeable future, possibly even for centuries. Recent research has provided new insights into the potential impacts of global warming on physiological function and temperature tolerance (Somero, 2012). Moreover, the impacts of climate change extend to various areas, such as alterations in freshwater stream and river flow patterns, heightened occurrence of hypoxic episodes, intensified storm activity, and shifts in ocean circulation and acidification. These developments will put the biomechanical performance of organisms to the test.

Biomechanics is a field that delves into the intricate workings of animal movement and structure, examining the underlying biological processes and materials involved. The environment plays a crucial role in shaping the qualities of ecosystems, leading to significant impacts (Somero, 2012). As storm activity and wave motion increase, intertidal communities can experience significant impacts. The biomechanical stress caused by increasing wave motion can have detrimental effects on the macroscopic algae and crustaceans found on rocky coasts. In some cases, these organisms may be harmed or even completely removed. Some organisms have biomaterial characteristics that enhance their ability to withstand intense physical impacts (Somero, 2012).

Changes in the surrounding environment, such as temperature fluctuations, the impact of wind waves, and ocean acidity, all work together to amplify their effects on individuals and the interactions between different species. Anthropogenic climate change is a significant selection pressure that has emerged in many ecosystems. Biomechanics has the potential to greatly enhance our understanding of the impacts of climate change. However, it has yet to fully focus on addressing the emerging challenges associated with the Anthropocene. It is crucial to emphasize the importance of biomechanics in addressing climate change at this particular moment (Somero, 2012).

It can be quite challenging to integrate the small-scale physiological and biomechanical characteristics of individuals with the large-scale biogeographic climatic factors. Biomechanics offers valuable insights into the effects of environmental stressors on an organism's movement and mechanical traits. Additionally, it offers predictive mechanistic models that significantly enhance our comprehension of the ecological impacts of climate change. In this discussion, we will explore the primary environmental changes associated with climate change and their potential impact on the biomechanics and motility of organisms, as outlined by Somero (2012).

10.3.1. Musculoskeletal Function and Movement

An animal's ability to move is influenced by various factors, including muscle, neuron, cardiovascular, and metabolic processes (Somero, 2012). Physiological systems can be influenced by various environmental factors, such as temperature, pH, oxygen levels, water flow, and wind speed. The impact of climate change on these characteristics can be observed in two distinct ways.

- Firstly, it leads to gradual alterations in the temperature and pH levels of the water.
- Secondly, it results in a higher occurrence of extreme weather events.

Environmental impacts on animal mobility can have far-reaching consequences for ecological networks, such as food web structures and the flow of energy across trophic levels (Somero, 2012) (Figure 10.2).

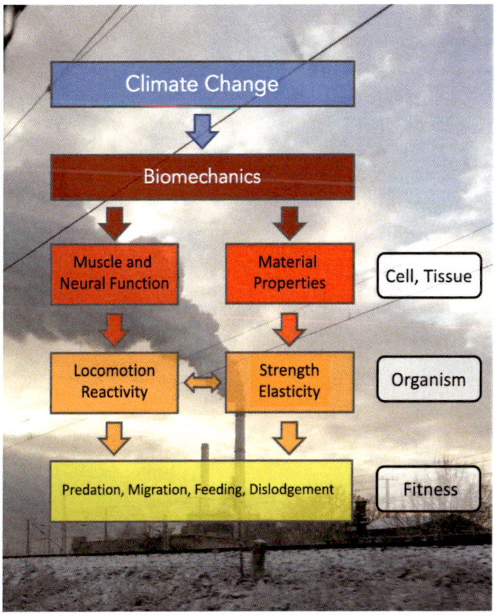

Figure 10.2. *The correlation between climate change and biomechanics, and the potential consequences for the fitness of organisms.*

Source: Domenici & Seebacher (2020).

The scale of environmental variability that affects performance and reproductive success depends on the mobility of individuals and species. As Somero (2012) explains, the reproductive success of migrating birds is influenced by a combination of local variation at breeding and overwintering grounds, as well as global variation caused by migration. The presence of different wind patterns and air currents has a significant impact on bird populations and their ability to survive. This is because these factors directly affect the biomechanics and energetics of their flight. Contrasting with mobile organisms, sessile creatures like intertidal mussels experience limited exposure to local environmental changes, despite the intricacies of their surroundings (Somero, 2012).

Mobile species such as fish and birds undergo extended periods of endurance training in response to varying climate conditions (Somero, 2012). Endurance exercise is influenced by muscle fatigue due to the biomechanical constraints on muscle function and the supply of metabolic energy. The mobility target may face challenges due to various environmental factors, such as fluctuations in temperatures and flows (such as streams, air, and ocean currents). These factors can also lead to increased transportation costs and decreased muscular power production (Somero, 2012). As we delve into the topic, it becomes evident that the reproductive success of migrating salmon is greatly influenced by factors such as warmer water and varying flow conditions. This, in turn, leads to a decrease in their overall reproductive success.

Climate change has led to the rise of ocean waves and temperatures. According to Somero (2012), it has been observed that higher wave heights can result in a more significant impact force, potentially leading to increased physiological strain on marine life. Temperature fluctuations have a significant impact on the contractile function of muscles, leading to potential variations in the effects of other factors. Intertidal creatures face a multitude of environmental factors, with wave activity being of utmost importance. Waves have the power to displace and eliminate individuals from the community (Somer, 2012). Adhesion strength to the substrate is a critical characteristic for the survival of intertidal creatures. As wave action increases, the snail's foot muscle needs to expend more energy to ensure that the animal remains in place. Furthermore, if the increase in wave action continues, a greater amount of force will need to be exerted for an extended duration. The endurance of an intertidal snail is found to be negatively affected by increasing water temperature. However, the snail's muscle tenacity is positively influenced by the presence of increased wave movement, resulting in a beneficial training effect. According to Somero (2012), the positive impact of increased tenacity in snails, which could help them withstand the strong forces of wave action, would be counteracted by the warming of the ocean.

It is undeniable that climate change and its consequences are inevitable. Enhanced understanding of the impact of climate change on existing ecosystems will lead to better management and conservation of natural resources and biodiversity (Somero, 2012). According to Somero (2012), the biomechanics of animals are mainly influenced by the combined effects of climatic change on material properties and muscle/neural functioning. The impact of these factors on the strength and ability to move of an organism could potentially influence its overall fitness. Ecological biomechanics, also known as ecomechanics, provides a comprehensive approach to studying the relationship between an organism's structure and function and its environment. This field of research explores how changes in an individual's physical characteristics are influenced by their surroundings and how these changes impact interactions within communities and populations (Somero, 2012).

10.4. ADAPTATION MECHANISMS IN ANIMALS

Animals need to adapt in order to survive and thrive in their environments. Which problems require resolution, such as low fertility and nutritional difficulties? Scientists studying animals and livestock producers face a number of crucial queries that demand answers (Fanta, 2017). Animal adaptability is influenced by various interrelated aspects, such as animal management and resources. For instance, these factors play a crucial role in determining how well animals can adapt to their environment. Animal adaptation encompasses not only the physical characteristics of the animal, but also the caregivers' capacity to adjust and optimize the available resources such as land, feed, water, and finances. When considering the adaptability of animals, it is crucial to thoroughly examine all factors that can either enhance or hinder their adaptability. Furthermore, it is crucial to consider both short- and long-term strategies for improving adaptation, as highlighted by Fanta (2017).

10.4.1. Morphological Adaptation

Examples of morphological adaptations include having slender legs, lighter skin, shorter and thinner hair, lighter hair color, higher sweat gland density, and less subcutaneous fat. The coat plays a crucial role in protecting against the harmful effects of direct sun radiation. As Fanta (2017) points out, cows living in tropical climates and sporting light-colored coats have the remarkable ability to reflect solar radiation. This unique adaptation serves as a protective shield, safeguarding the animal from the potentially harmful effects of the sun's rays.

According to Fanta (2017), cattle with darker coats have the ability to absorb more sunlight, resulting in increased heat production. Cattle acclimated to dry environments benefit from the slick hair gene, as it enables them to effectively dissipate heat. This gene results in smooth, short, and thin hair. Animals are able to thrive in hot regions due to their ability to sweat. The rate at which cattle sweat and the density of their sweat glands are closely linked to their ability to tolerate heat. Animals living in warm climates have sweat glands that are larger in diameter, volume, perimeter, and density. In addition, it has been observed that cattle from tropical regions generally have smaller bodies when compared to breeds from temperate locations (Fanta, 2017).

Sheep with light-colored coats have the ability to reflect a greater amount of solar radiation compared to sheep with dense, black, or shaggy coats. According to Fanta (2017), light-colored coats are known for their sleek and lustrous appearance. In addition, sheep with carpet wool are able to enhance their protection against solar radiation by promoting cutaneous heat dissipation. According to Fanta (2017), sheep with longer, thicker, and darker coats tend to experience more stress and exhibit higher rectal temperatures and sweating rates in tropical climates, when compared to their white-haired counterparts. Goats possess remarkable adaptability to thrive in desert environments. Goats possess unique physiological traits that give them an advantage over other ruminant species in challenging environmental conditions. Surviving in extreme climates is made possible for these creatures due to their small size, unique fleece structure, and remarkable digestive efficiency (Fanta, 2017). Moreover, it is worth noting that dwarf goats exhibit a higher level of adaptability in arid regions compared to other breeds. This can be attributed to their light-colored coats and petite, upright, forward-facing ears. Goats inhabiting arid regions possess lengthy hair and rough fleeces, enabling them to regulate their body temperature in the scorching daytime and freezing nighttime conditions. Furthermore, in order to withstand harsh winter conditions, goats in temperate regions develop a dual-layered coat consisting of short, fine fibers and long, coarse fibers (Fanta, 2017).

10.4.2. Behavioral Adaptation

Animals undergo behavioral reactions to effectively acclimate when exposed to high heat loads. Behavioral responses observed in heat-stressed ruminants include seeking shade, reduced feed intake, increased water intake and drinking frequency, prolonged standing time, reduced lying time, and decreased frequency of feces and urine (Curtis et al., 2017). One of the immediate behavioral responses of an animal when it becomes overheated is to actively search for shade. On clear days with temperatures exceeding 21°C, dairy cattle typically prefer to find shade. As the air temperature and sun radiation rise, cows tend to seek more shade, often spending over 10 hours a day in shaded areas. Shade can have a significant impact on grazing time, ultimately affecting milk production and growth. Sheep, despite their typically robust nature, will seek shelter when faced with elevated temperatures.

One strategy that animals employ during the summer to lower their metabolic heat output is to reduce their meal intake. Various studies have demonstrated that heat exposure can lead to reduced feed consumption in cattle, sheep, and goats (Curtis et al., 2017). In a study conducted by Curtis et al. (2017), it was found that behavioral research showed differences in the grazing habits of cattle when they were managed in hot weather. The study found that during the day, the grazing duration was reduced, while during the night, it was increased. During periods of high temperatures, ruminant cattle tend to increase their water intake and do so more frequently.

According to Chedid et al. (2014), it was found that desert sheep have the ability to concentrate their urine during periods of acute heat load in order to compensate for the increased water loss. Similarly, Brscic et al. (2007) conducted research on heat-stressed cattle and observed a decrease in the frequency of urination. According to Brscic et al. (2007), the duration of standing and lying down can be influenced by high heat load. According to research conducted by Chedid et al., in 2014, sheep and cattle tend to adjust their positions and seek shade in response to the heat, in order to protect themselves from the sun and the hot ground.

10.4.3. Physiological Adaptation

Physiological flexibility is a crucial response mechanism that enables animals to endure and survive under extreme heat conditions. When animals are exposed to heat, their internal heat load decreases because their bodies are able to dissipate excess heat to the surroundings. In addition, various physiological reactions, such as increased skin temperature, pulse rate, respiration rate, rectal temperature, and sweating rate, play a role in dissipating excess body heat (Chauhan et al., 2014). Throughout the day, there are noticeable diurnal fluctuations in physiological responses, whereas during the night, the levels remain constant. Animals that experience a decrease in body heat during the nighttime hours possess a remarkable ability to endure the heightened temperatures that they encounter during the daytime. Rectal temperature and respiration rate are considered the most reliable indicators of heat stress in various ruminant species, as suggested by Chauhan et al. in 2014.

10.4.4. Blood Biochemical Adaptation

Heat stress can lead to changes in blood biochemical characteristics. The rise in hemoglobin and packed cell volume in cattle is a result of heat stress. These changes are believed to be the gradual development of adaptive characteristics in cattle. In addition, several hormones play a crucial role in regulating the homeothermy mechanism in ruminant animals. Animals adapt to increasing ambient temperatures by reducing the secretion of thyroid hormones, which play a crucial role in regulating metabolic processes and body heat generation. In ruminant animals, cortisol is considered the primary biochemical marker for heat stress (Abramowicz et al., 2019). When ruminants experience heat stress, their cortisol levels tend to increase significantly, which is a clear sign of stress. Superoxide dismutase and glutathione peroxidase serve as indicators of oxidative stress in sheep and cattle, particularly in situations of elevated heat exposure. According to a study conducted by Abramowicz et al. in 2019, there has been evidence of an increase in the content of antioxidants in dairy cattle and sheep.

Livestock plays a crucial role in contributing to a substantial portion of the global food production. Animals play a crucial role in providing high-quality food, particularly for farmers in impoverished nations who heavily depend on them for their livelihood. One of the primary objectives of the agriculture sector is to sustain cattle production in the midst of climate change challenges. In order to mitigate the adverse impacts of climate change on livestock, it is imperative to adopt interdisciplinary approaches that encompass animal breeding, feeding, housing, and health. Understanding how livestock responds to its surroundings and making changes to nutritional and environmental management tactics are essential for improving animal comfort and performance (Abramowicz et al., 2019). One of the significant challenges in developing a climate change adaptation plan is the inherent uncertainty that needs to be addressed.

10.5. ROLE OF NUTRITIONAL STRATEGIES IN ENVIRONMENTAL STRESS

Animals need to consume well-balanced diets that contain all the necessary nutrients for their growth, reproduction, and to develop a strong immune system to protect them from diseases in challenging environments (Parker et al., 2009). In this section, we will explore several environmental factors that have a detrimental effect on animal performance. These factors include heat stress, cold stress, disease outbreak, and high altitude.

Heat stress and nutrition are important factors to consider when it comes to maintaining optimal health. The impact of heat stress on the body can be significant, and it is crucial to understand how nutrition can play a role in mitigating its effects. By ensuring a well-balanced diet and proper hydration, individuals can better equip themselves to cope with heat stress and support their overall well-being. Here are some feeding techniques that can help with climate adaptation:

1. **Protein Nutrition:**Understanding the importance of protein in our diet can help us make informed choices when it comes to our nutritional intake. By ensuring we consume an adequate amount of protein. A study conducted by Parker et al. in 2009 examined the effects of protein supplementation on Holstein cows during periods of natural heat stress. When implementing the high-protein diet, there was a notable 11% rise in dry matter consumption, along with a 4.3% increase in fat-corrected milk (FCM). Excessive protein intake should be minimized as it can negatively impact energy utilization. This is due to the fact that the body requires additional energy to eliminate excess protein through the urea cycle (Parker et al., 2009).
2. **Fat Nutrition:**In hotter regions, it is essential to include a moderate amount of fat in the diet, specifically ranging from 3% to 8%, in

order to prevent heat stress (Parker et al., 2009). Consequently, the increase in heat associated with tissue metabolism and fermentation in the digestive system is reduced, resulting in a decrease in the body's total heat load. Feed additives such as whole cottonseed, soybean oil, or hydrogenated vegetable fat are highly valued for their ability to reduce heat load and increase milk supply and milk fat percentage (Parker et al., 2009).

3. **Mineral Supplementation:**Supplementing with minerals is highly recommended in cases of heat stress, particularly in high-yielding dairy cows. Potassium (K) plays a crucial role as it is necessary to replenish the losses caused by perspiration and milk. During heat stress, there is a decrease in blood aldosterone levels, which leads to an increase in urinary sodium (Na) losses. It is important to consider the use of dietary supplements to counteract these losses, as suggested by Parker et al. (2009).
4. **Using Cold Water:** According to the findings, when lactating cows were provided with water chilled to 10°C instead of non-chilled water at 27°C, several notable differences were observed. These included lower respiration rates (70 vs. 81/minute), lower rectal temperatures (39 vs. 40°C), and higher milk output (26 vs. 24 kg/d) (Parker et al., 2009). Using cooled water may not be necessary if the cows have access to fans or sprinklers.

In colder climates, pregnant animals may be more prone to pregnancy ketosis and hypomagnesemia due to changes in their endocrine and metabolic processes. In a study conducted by Parker et al. in 2009, it was found that the maintenance energy needs of cattle increase by 0.91% for every degree below 20°C. This decrease in temperature negatively affects the efficiency of energy consumption for growth, lactation, and other biological processes.

Here are some helpful hints for reducing the negative effects of cold stress through feeding techniques:

- In order to ensure the well-being of cows in the final trimester, it becomes necessary to increase their grain intake when the effective temperature falls below the lower critical limit.
- It would be beneficial to increase the amount of grain and hay provided. It is not advisable to freeze wet feeds.
- It is crucial to ensure that cows have constant access to an adequate water supply. Limiting access to water will result in a decrease in cows' consumption of feed and make it more challenging for them to meet their energy requirements. The intake of drinking water is

greatly limited due to frozen troughs and the presence of extremely cold water.

- In cold weather, the maturity and composition of feed ingredients can be impacted. Hence, it is necessary to make appropriate modifications to the feed formula for animals experiencing stress due to cold weather.

Proper conservation and the application of proven chemical and physical treatment techniques are crucial for enhancing the digestibility and utilization of feeds at high elevations, where their nutritional value is reduced. Incorporating antioxidant supplements, such as vitamin E, into diets is recommended (Parker et al., 2009). Through the reduction of oxidative stress commonly experienced at high elevations and the optimization of feed intake and nutrient utilization, various benefits are achieved, including improved health, increased body weight gain, and enhanced survival rates. These factors ultimately contribute to enhanced overall performance.

In high elevations, the presence of lower oxygen levels and atmospheric pressure necessitates the provision of adequate iron levels (Parker et al., 2009). Iron plays a crucial role in enhancing performance and aiding in high-altitude adaptation. It does so by boosting the blood's ability to transport oxygen and facilitating its consumption by the cells. It is worth noting that conventional cereals and legumes used in feeding contain non-heme iron, which is not easily absorbed and utilized by animals. In order to ensure that animals raised at high altitudes receive the proper nutrition, it is necessary to make appropriate adjustments to their diets. This includes incorporating heme-iron sources derived from animals, while also maintaining the ideal balance of protein and energy (Parker et al., 2009).

10.6. CONSERVATION AND BIODIVERSITY IN ANIMAL POPULATIONS

Biodiversity refers to the variety of living organisms in a specific area or on Earth as a whole. It encompasses the diversity of species, genetic variation within species, and the variety of ecosystems. Biodiversity is crucial for the functioning of ecosystems and provides numerous benefits to human society. It plays a vital role in maintaining ecosystem stability, supporting food production, regulating climate, providing medicines, and enhancing overall resilience to environmental changes.

Biodiversity, as defined by Frankham et al. (2012), encompasses the range of life forms found on Earth, including both the genetic diversity within a population and the biological diversity within an ecosystem. This encompasses both the overall abundance of living organisms in a specific area and the diversity of variations among those organisms. While some individuals inherently appreciate the biodiversity of certain organisms, such as bacteria, insects, and plants, others do not share the same sentiment. However, these often ignored species perform vital functions in maintaining environmental equilibrium (Frankham et al., 2012). The absence of diversity at any level of life can lead to the collapse of an ecosystem.

Biodiversity is crucial for the well-being of life on Earth. The reason for this is that ecosystems with a greater level of biodiversity, including a variety of species and individuals, tend to have higher ecological stability. This increased stability enhances the ability of populations within these ecosystems to withstand various disturbances, such as disease outbreaks and climate change.

Researchers commonly study three tiers of biodiversity: genetic diversity, species diversity, and ecosystem diversity. Conservation efforts usually prioritize the preservation of species diversity. Species diversity, although more manageable in scale, can offer insights into the well-being of genetic and ecosystem diversity, both of which are highly important. A lack of species diversity in an ecosystem can affect both the genetic diversity of the species within it and the overall status of the ecosystem in relation to other ecosystems. A biodiversity index is a quantitative tool used to measure species biodiversity. A basic biodiversity index can be calculated by dividing the species richness (total number of species) by the species evenness (total number of individuals) in a given area (Frankham et al., 2012).

Scientists have found that the rate of biodiversity loss is unprecedented, as evidenced by indicators such as the biodiversity index. In addition, the loss of biodiversity is happening in certain equatorial regions, referred to as biodiversity hotspots, at a more rapid pace compared to other regions. Human activities are primarily responsible for the main causes of biodiversity decline, including climate change. Consequently, a destructive cycle of harm and destruction occurs due to the continuous decline of biodiversity in ecosystems worldwide.

10.6.1. Preservation

While some individuals inherently appreciate the diversity of certain organisms, such as bacteria, insects, and plants, others do not share the same sentiment. However, these often ignored species perform essential functions in maintaining environmental equilibrium. The absence of diversity at any level of life can lead to the collapse of an ecosystem. Biodiversity is crucial for the well-being of life on Earth (Frankham et al., 2012). The presence of greater biodiversity among individuals, animals, and ecosystems within populations enhances their resilience to various disruptions, such as disease outbreaks and climate change.

Researchers often explore three levels of biodiversity: genetic diversity, species diversity, and ecosystem diversity. Conservation efforts typically focus on preserving species diversity. While species diversity is easier to manage, it can provide valuable insights into the health of both genetic and ecosystem diversity, which are both very important. A lack of species diversity in an ecosystem can affect the genetic diversity of the species within it and also its overall position relative to other ecosystems (Frankham et al., 2012).

A biodiversity index is a quantitative tool utilized to measure species biodiversity. A basic biodiversity index can be calculated by dividing the species richness (total number of species) by the species evenness (total number of individuals) in a given area.

Scientists have discovered that the rate of biodiversity loss is at an unprecedented level, as shown by indicators like the biodiversity index. Furthermore, the loss of biodiversity is happening at accelerated rates in certain equatorial regions, commonly referred to as biodiversity hotspots, compared to other regions. Human activities are primarily responsible for the main causes of biodiversity decline, including climate change. Consequently, the continuous decline of biodiversity in ecosystems worldwide leads to a harmful cycle of loss and damage (Frankham et al., 2012).

10.6.2. Criteria for Conservation: Who and Why Are Selected?

Conservation efforts aim to mitigate the negative impact of human activities on the earth by protecting species and their habitats, thereby promoting the recovery of ecosystems and population levels. Conservation efforts typically fall into two main categories: either they prevent deforestation and other harmful activities from affecting ecosystems, or they promote the growth and protection of populations to aid in the recovery of injured populations. Typically, these efforts are categorized as either ex-situ (preservation of biodiversity elements in locations other than their original environments, like zoos, seed banks, and botanical gardens) or in-situ (preservation of habitats, species, and ecosystems in their natural locations) (Frankham et al., 2012).

Resource constraints, such as limited funds, available land, and workforce, necessitate the prioritization of certain locations or species for conservation over others. Hotspots of diversity are a widely used approach for selecting priority locations for conservation. Due to their significant levels of species richness, rates of endemism (uniqueness to a specific region), and concerns regarding survival, these "hotspots" are areas that require urgent conservation efforts. Currently, there are 36 hotspots that have been officially designated worldwide.

The equator is the primary location for these biogeographic zones, and hotspots have experienced a loss of approximately 85% of their habitat.

The IUCN Red List is a metric utilized to evaluate the status of individual species (Frankham et al., 2012). The International Union for Conservation of Nature (IUCN) is a global organization committed to safeguarding wildlife and advocating for the sustainable utilization of natural resources. Its Red List is renowned for being the most comprehensive inventory of the global status of biological species conservation. They assert to have assessed 42,100 species that are in danger of becoming extinct, however, a 2019 report from the United Nations indicates that the total number of endangered species is 1 million. The discrepancy arises from the limited number of species assessed by the IUCN (Frankham et al., 2012).

10.7. PROSPECTS FOR FUTURE RESEARCH IN THE FIELD OF ENVIRONMENTAL ANIMAL BIOCHEMISTRY

The field of animal clinical biochemistry will undoubtedly be influenced by the changing landscape in the medical biochemistry field due to new and coordinated requirements for cost containment. The animal clinical biochemist must possess thorough preparation and a keen awareness in order to effectively respond to these demands and the resulting changes. The animal clinical biochemist is a scientist who has the unique ability to understand and navigate across different disciplines in order to grasp the subtle differences between species. Consequently, the animal clinical biochemist is required to remain up-to-date with emerging trends in the field and actively contribute to its growth (Bell, 2007).

This implies that individuals employed in the industry have a fundamental responsibility to conduct research and distribute new information. This allows the animal clinical biochemist to assume the role of a principal investigator or independent scientist for a range of projects related to animals. This implies that animal clinical biochemists must enhance their comprehension of clinical medicine in order to effectively communicate using medical terminology (Bell, 2007).

There are numerous emerging opportunities that animal clinical biochemists can exploit to broaden the scope and diversity of their work. The mentioned topics encompass point-of-care systems, innovative approaches to treating diabetes, autoimmune disorders, thyroid conditions, and tumor indicators (Bell, 2007). The list of additional categories is extensive and encompasses DNA probes for hereditary diseases, indicators of infectious diseases, indicators of nutritional illnesses or deficiencies, and indicators of inflammation. These fields are primed for novel viewpoints and scientific breakthroughs; all that is required is the innovative vigor of the animal clinical biochemist (Bell, 2007).

Ultimately, we encounter challenges and solutions in the intricate interplay of biology and the environment. Biochemistry serves as the foundation for human activities that have significant effects on the world, including urbanization, deforestation, and agricultural practices. Our actions, such as burning fossil fuels and modifying the atmosphere, have a significant impact on the intricate processes of life. Additionally, our actions can also influence the oceans through nitrogen runoff. Nevertheless, biochemistry holds great potential for the future of sustainable coexistence as it elucidates the mechanisms by which organisms adapt, purify, and regenerate themselves. Put simply, the molecular strands that connect all living organisms on Earth can be found in both forest ecosystems and bustling urban areas.

Bibliography

1. Abramowicz, B., Kurek, Ł., & Lutnicki, K. (2019). Hematology in the early diagnosis of cattle diseases: A review. *Veterinarski Arhiv*, *89*, 579–590.
2. Aehle, W. (2007). *Enzymes in Industry: Production and Applications*. John Wiley & Sons.
3. Alvares, C. (2024). 17 enigmatic facts about Michaelis-Menten equation. *Facts.net*. https://facts.net/science/chemistry/17-enigmatic-facts-about-michaelis-menten-equation/ (accessed on 20 August 2024).
4. Amine, A., Mohammadi, H., Bourais, I., & Palleschi, G. (2006). Enzyme inhibition-based biosensors for food safety and environmental monitoring. *Biosensors and Bioelectronics, 21*(8), 1405–1423.
5. Aminzadeh-Gohari, S., Feichtinger, R. G., & Kofler, B. (2019). Energy metabolism and metabolic targeting of neuroblastoma. In *Neuroblastoma* (pp. 113–132). Academic Press.
6. Atta, A. H., Atta, S. A., Nasr, S. M., & Mouneir, S. M. (2022). Current perspective on veterinary drug and chemical residues in food of animal origin. *Environmental Science and Pollution Research International, 29*(11), 15282–15302. https://doi.org/10.1007/s11356-021-18239-y.
7. Bakke, A. M., Glover, C., & Krogdahl, Å. (2010). Feeding, digestion and absorption of nutrients. In *Fish Physiology* (Vol. 30, pp. 57–110). Academic Press.
8. Bartee, L., Shriner, W., & Creech, C. (2017). Feedback inhibition in metabolic pathways. *Pressbooks*. Retrieved from: https://openoregon.pressbooks.pub/mhccmajorsbio/chapter/6-7-feedback-inhibition-in-metabolic-pathways/ (accessed on 20 August 2024).
9. Bechet, D. D., Tassa, A., Taillandier, D., Combaret, L., & Attaix, D.

(2005). Lysosomal proteolysis in skeletal muscle. *International Journal of Biochemistry & Cell Biology, 37*(10), 2098–2114. https://doi.org/10.1016/j.biocel.2005.02.029.

10. Bell, A. M. (2007). Future directions in behavioral syndromes research. *Proceedings of the Royal Society B: Biological Sciences, 274*(1611), 755–761.
11. BeMiller, J. N. (2018). *Carbohydrate Chemistry for Food Scientists*. Elsevier.
12. Bert, B., Chmielewska, J., Hensel, A., Grune, B., & Schönfelder, G. (2016). The animal experimentation quandary: Stuck between legislation and scientific freedom. *EMBO Reports, 17*(6), 790–792.
13. Birsoy, K., Festuccia, W. T., & Laplante, M. (2013). A comparative perspective on lipid storage in animals. *Journal of Cell Science, 126*(7), 1541–1552.
14. Bonora, M., Patergnani, S., Rimessi, A., De Marchi, E., Suski, J. M., Bononi, A., & Pinton, P. (2012). ATP synthesis and storage. *Purinergic Signaling, 8*(3), 343–357.
15. Braverman, N. E., & Moser, A. B. (2012). Functions of plasmalogen lipids in health and disease. *Biochimica et Biophysica Acta (BBA)–Molecular Basis of Disease, 1822*(9), 1442–1452.
16. Brieger, K., Schiavone, S., Miller, F. J., Jr., & Krause, K. H. (2012). Reactive oxygen species: From health to disease. *Swiss Medical Weekly, 142*(3334), w13659.
17. Brscic, M., Gottardo, F., Mazzenga, A., & Cozzi, G. (2007). Behavioral response to different climatic conditions of beef cattle in intensive rearing systems. *Poljoprivreda, 13*(1), 103–106.
18. Buijtels, H. K. S. G. J., & Meij, B. P. (2009). Endocrine diseases in animals. *Hormone Research in Paediatrics, 71*(1), 144–147.
19. BYJUS. (2023). Nutrition in animals: Human digestive system, types of nutrition in animals. Retrieved from: https://byjus.com/cbse/nutrition/ (accessed on 20 August 2024).
20. Chauhan, S. S., Celi, P., Leury, B. J., Clarke, I. J., & Dunshea, F. R. (2014). Dietary antioxidants at supranutritional doses improve oxidative status and reduce the negative effects of heat stress in sheep. *Journal of Animal Science, 92*(8), 3364–3374.
21. Chedid, M., Jaber, L. S., Giger-Reverdin, S., Duvaux-Ponter, C., & Hamadeh, S. K. (2014). Water stress in sheep raised under arid conditions. *Canadian Journal of Animal Science, 94*(2), 243–257.

22. Choct, M. (2006). Enzymes for the feed industry: Past, present and future. *World's Poultry Science Journal, 62*(1), 5–16.
23. Copeland, R. A. (2023). *Enzymes: A Practical Introduction to Structure, Mechanism, and Data Analysis*. John Wiley & Sons.
24. Cray, C., Zaias, J., & Altman, N. H. (2009). Acute phase response in animals: A review. *Comparative Medicine, 59*(6), 517–526.
25. Crowe, M. A., Hostens, M., & Opsomer, G. (2018). Reproductive management in dairy cows: The future. *Irish Veterinary Journal, 71*(1), 1–13.
26. Curtis, A. K., Scharf, B., Eichen, P. A., & Spiers, D. E. (2017). Relationships between ambient conditions, thermal status, and feed intake of cattle during summer heat stress with access to shade. *Journal of Thermal Biology, 63*, 104–111.
27. Daly, J., Smith, H., McGrice, H. A., Kind, K. L., & van Wettere, W. H. (2020). Towards improving the outcomes of assisted reproductive technologies of cattle and sheep, with particular focus on recipient management. *Animals, 10*(2), 293.
28. Dashty, M. (2013). A quick look at biochemistry: Carbohydrate metabolism. *Clinical Biochemistry, 46*(15), 1339–1352.
29. Devolli, A., Stafasani, M., Shahinasi, E., & Hamiti, H. (2021). Determination of vitamin C content in commercial fruit juices by volumetric and spectrophotometric methods. *ResearchGate*. https://www.researchgate.net/publication/350739712_DETERMINATION_OF_VITAMIN_C_CONTENT_IN_COMMERCIAL_FRUIT_JUICES_BY_VOLUMETRIC_AND_SPECTROPHOTOMETRIC_METHODS (accessed on 20 August 2024).
30. Dhakal, S., Subhan, M., Fraser, J. M., Gardiner, K., & Macreadie, I. (2019). Simvastatin efficiently reduces levels of Alzheimer's amyloid beta in yeast. *International Journal of Molecular Sciences, 20*(14), 3531. https://doi.org/10.3390/ijms20143531.
31. Domenici, P., & Seebacher, F. (2020). The impacts of climate change on the biomechanics of animals. *Conservation Physiology, 8*(1). https://doi.org/10.1093/conphys/coz102.
32. Egom, E. E. A., & Hafeez, H. (2016). Biochemistry of statins. In *Advances in Clinical Chemistry* (Vol. 73, pp. 127–168). Elsevier.
33. Ehtisham, M., Wani, F., Wani, I., Kaur, P., & Nissar, S. (2016). Polymerase chain reaction (PCR): Back to basics. *Indian Journal of Contemporary Dentistry, 4*(2), 30.

34. Fanta, M. (2017). Physiological adaptation of Holstein Frisian dairy cattle in Ethiopia: Review article. *Journal of Biology, Agriculture and Healthcare, 7*(67–78).
35. Feingold, K. R. (2024). Introduction to lipids and lipoproteins. In *Endotext* [Internet]. South Dartmouth (MA): MDText.com, Inc. https://www.ncbi.nlm.nih.gov/sites/books/NBK305896/ (accessed on 20 August 2024).
36. Firuzi, O., Miri, R., Tavakkoli, M., & Saso, L. (2011). Antioxidant therapy: Current status and future prospects. *Current Medicinal Chemistry, 18*(25), 3871–3888.
37. Frankham, R., Ballou, J. D., Dudash, M. R., Eldridge, M. D., Fenster, C. B., Lacy, R. C., & Ryder, O. A. (2012). Implications of different species concepts for conserving biodiversity. *Biological Conservation, 153*, 25–31.
38. Gardner, G. E., McGilchrist, P., & Pethick, D. W. (2014). Ruminant glycogen metabolism. *Animal Production Science, 54*(10), 1575–1583.
39. Gebert, L. F., & MacRae, I. J. (2019). Regulation of microRNA function in animals. *Nature Reviews Molecular Cell Biology, 20*(1), 21–37.
40. Gérard, P. (2013). Metabolism of cholesterol and bile acids by the gut microbiota. *Pathogens, 3*(1), 14–24.
41. Haida, Z., & Hakiman, M. (2019). A comprehensive review on the determination of enzymatic assay and nonenzymatic antioxidant activities. *Food Science & Nutrition, 7*(5), 1555–1563.
42. Haneef, J. (2023). Classification of enzymes: 6 important classes of enzymes and their biochemistry with examples. *Biology Exams 4 U.* https://www.biologyexams4u.com/2023/05/classification-of-enzymes-6-important.html#google_vignette (accessed on 20 August 2024).
43. Heindel, J. J., Blumberg, B., Cave, M., Machtinger, R., Mantovani, A., Mendez, M. A., & Vom Saal, F. (2017). Metabolism disrupting chemicals and metabolic disorders. *Reproductive Toxicology, 68*, 3–33.
44. Hill, J. O., Wyatt, H. R., & Peters, J. C. (2012). Energy balance and obesity. *Circulation, 126*(1), 126–132.
45. Hu, Y., Cheng, H., & Tao, S. (2017). Environmental and human health challenges of industrial livestock and poultry farming in China and their mitigation. *Environment International, 107*, 111–130.
46. Ibtisham, F., Nawab, A., Li, G., Xiao, M., An, L., & Naseer, G. (2018). Effect of nutrition on reproductive efficiency of dairy animals. *Medycyna Weterynaryjna, 74*(1), 6025–2018. https://doi.org/10.21521/mw.6025.
47. Ighodaro, O., & Akinloye, O. (2018). First line defense antioxidants-superoxide dismutase (SOD), catalase (CAT) and glutathione peroxidase

(GPX): Their fundamental role in the entire antioxidant defense grid. *Alexandria Journal of Medicine, 54*(4), 287–293.

48. JaypeeDigital. (2015). *Clinical Biochemistry*. Retrieved from: https://www.jaypeedigital.com/book/9789351529255/chapter/ch27 (accessed on 20 August 2024).
49. Jordan, S. D., Könner, A. C., & Brüning, J. C. (2010). Sensing the fuels: Glucose and lipid signaling in the CNS controlling energy homeostasis. *Cellular and Molecular Life Sciences, 67*(19), 3255–3273. https://doi.org/10.1007/s00018-010-0414-7.
50. Judge, A., Dodd, A., & Michael, D. (2020). Metabolism. *Essays in Biochemistry, 64*(1–17). https://doi.org/10.1042/EBC20190041.
51. Kaneko, J. (2000). A century of animal clinical biochemistry: Growth, maturity and visions for the future. *Clinical Biochemistry*.
52. Kaneko, J. J., Harvey, J. W., & Bruss, M. L. (2008). *Clinical Biochemistry of Domestic Animals*. Academic Press.
53. Khan Academy. (2024). Enzyme regulation. *Khan Academy*. Retrieved from: https://www.khanacademy.org/science/ap-biology/cellular-energetics/environmental-impacts-on-enzyme-function/a/enzyme-regulation (accessed on 20 August 2024).
54. Knudsen, K. E. B., Hedemann, M. S., & Lærke, H. N. (2012). The role of carbohydrates in intestinal health of pigs. *Animal Feed Science and Technology, 173*(1–2), 41–53.
55. Krauss, G. (2006). *Biochemistry of Signal Transduction and Regulation*. John Wiley & Sons.
56. Kumar, A., Kumari, C., Mochan, S., Kulandhasamy, M., Sesham, K., & Sharma, V. K. (2022). Endocrine system. In *Encyclopedia of Animal Cognition and Behavior* (pp. 2311–2337). Cham: Springer International Publishing.
57. Larigot, L., Juricek, L., Dairou, J., & Coumoul, X. (2018). AhR signaling pathways and regulatory functions. *Biochimie Open, 7*, 1–9.
58. Li, J. M., Li, L. Y., Qin, X., Ning, L. J., Lu, D. L., Li, D. L., & Du, Z. Y. (2017). Systemic regulation of L-carnitine in nutritional metabolism in zebrafish, *Danio rerio*. *Scientific Reports, 7*(1), 40815.
59. Luisa, B. G. (2012). *Cellular Energy Metabolism and Its Regulation*. Elsevier.
60. Lykkesfeldt, J., & Svendsen, O. (2007). Oxidants and antioxidants in disease: Oxidative stress in farm animals. *The Veterinary Journal, 173*(3), 502–511.

61. Martínez-Reyes, I., & Chandel, N. S. (2020). Mitochondrial TCA cycle metabolites control physiology and disease. *Nature Communications, 11*(1). https://doi.org/10.1038/s41467-019-13668-3.
62. Maurya, N. (2018). Enzyme inhibition (competitive, non-competitive and uncompetitive). *New.* https://medicoapps.org/m-enzyme-inhibition-competitive-non-competitive-and-uncompetitive/ (accessed on 20 August 2024).
63. Mbiza, J. (2020). Difference between ruminant and non-ruminant animals. *Pediaa.com.* https://www.pinterest.com/pin/682717624748461228/ (accessed on 20 August 2024).
64. MetaFilter. (2006). Lineweaver-burk woes. Retrieved from: https://ask.metafilter.com/32682/Lineweaverburk-woes (accessed on 20 August 2024).
65. Michal, G., & Schomburg, D. (2012). *Biochemical Pathways: An Atlas of Biochemistry and Molecular Biology*. John Wiley & Sons.
66. Miyamoto, T., & Amrein, H. (2017). Gluconeogenesis: An ancient biochemical pathway with a new twist. *Fly, 11*(3), 218–223.
67. Mizushima, N., & Klionsky, D. J. (2007). Protein turnover via autophagy: Implications for metabolism. *Annual Review of Nutrition, 27*(1), 19–40.
68. Modric, J. (2015). Disaccharide definition, structure, list, food examples. *Nutrients Review.* https://www.nutrientsreview.com/carbs/disaccharides.html (accessed on 20 August 2024).
69. Modric, J. (2016a). Monosaccharides (simple sugars) definition, list, examples of foods. *Nutrients Review.* https://www.nutrientsreview.com/carbs/monosaccharides-simple-sugars.html (accessed on 20 August 2024).
70. Modric, J. (2016b). Polysaccharides definition, list, functions, food examples. *Nutrients Review.* https://www.nutrientsreview.com/carbs/polysaccharides.html (accessed on 20 August 2024).
71. Modric, J. (2017). Oligosaccharides definition, benefits, food examples. *Nutrients Review.* https://www.nutrientsreview.com/carbs/oligosaccharides.html (accessed on 20 August 2024).
72. Morrison, C. D., & Laeger, T. (2015). Protein-dependent regulation of feeding and metabolism. *Trends in Endocrinology & Metabolism, 26*(5), 256–262.
73. Muley, V. (2020). What is ruminant? Ruminant digestive system. *Indian Cattle.* https://indiancattle.com/digestive-system-of-ruminant/ (accessed on 20 August 2024).
74. Mullur, R., Liu, Y. Y., & Brent, G. A. (2014). Thyroid hormone regulation of metabolism. *Physiological Reviews*.

75. Nagansurkar, S. (2021). Protein synthesis. *Simple Pharma Notes*. https://www.simplepharmanotes.com/2020/04/protein-synthesis.html (accessed on 20 August 2024).

76. Nakamura, M. T., Yudell, B. E., & Loor, J. J. (2014). Regulation of energy metabolism by long-chain fatty acids. *Progress in Lipid Research, 53*, 124–144.

77. Nauck, M. A., & Meier, J. J. (2018). Incretin hormones: Their role in health and disease. *Diabetes, Obesity and Metabolism, 20*, 5–21.

78. News-Medical. (2020). A double-edged sword: Reactive oxygen species and how they are detected. Retrieved from: https://www.news-medical.net/whitepaper/20190603/A-double-edged-sword-reactive-oxygen-species-and-how-they-are-detected.aspx (accessed on 20 August 2024).

79. Nolan, J. (2007). *Nutrient Requirements of Domesticated Ruminants*. CSIRO Publishing.

80. Ojha, B. K., Singh, P. K., & Shrivastava, N. (2019). Enzymes in the animal feed industry. In *Enzymes in Food Biotechnology* (pp. 93–109). Academic Press.

81. Parker, K. L., Barboza, P. S., & Gillingham, M. P. (2009). Nutrition integrates environmental responses of ungulates. *Functional Ecology, 23*(1), 57–69.

82. Phillips, M. C. (2014). Apolipoprotein E isoforms and lipoprotein metabolism. *IUBMB Life, 66*(9), 616–623.

83. Pillai, R. S., Bhattacharyya, S. N., & Filipowicz, W. (2007). Repression of protein synthesis by miRNAs: How many mechanisms? *Trends in Cell Biology, 17*(3), 118–126.

84. Ponnampalam, E. N., Kiani, A., Santhiravel, S., Holman, B. W. B., Lauridsen, C., & Dunshea, F. R. (2022). The importance of dietary antioxidants on oxidative stress, meat and milk production, and their preservative aspects in farm animals: Antioxidant action, animal health, and product quality—Invited review. *Animals, 12*(23), 3279. https://doi.org/10.3390/ani12233279.

85. Pressbooks. (2020). The atom. Retrieved from: https://pressbooks.oer.hawaii.edu/humannutrition2/chapter/8-the-atom/ (accessed on 20 August 2024).

86. Rogatzki, M. J., Ferguson, B. S., Goodwin, M. L., & Gladden, L. B. (2015). Lactate is always the end product of glycolysis. *Frontiers in Neuroscience, 9*, 22.

87. Ross, A. C., Caballero, B., Cousins, R. J., & Tucker, K. L. (2020). *Modern Nutrition in Health and Disease*. Jones & Bartlett Learning.

88. Saha, S. K., & Pathak, N. N. (2021). *Fundamentals of Animal Nutrition* (pp. 219–246). Springer.

89. Santos, S., Rotta, P. P., Costa, E., Silva, L. F., Menezes, A. C. B., Pina, D. S., & Valadares Filho, S. C. (2016). Protein ruminal degradation of feeds and microbial protein synthesis. *Nutrient Requirements of Zebu and Crossbred Cattle*, 43.

90. Sarvasmaa, A. S. (2007). Sleep deprivation and brain energy metabolism: *In vivo* studies in rats and humans. ResearchGate. https://www.researchgate.net/publication/47931698_Sleep_deprivation_and_brain_energy_metabolism_in_vivo_studies_in_rats_and_humans (accessed on 20 August 2024).

91. Schug, T. T., Johnson, A. F., Birnbaum, L. S., Colborn, T., Guillette Jr, L. J., Crews, D. P., & Heindel, J. J. (2016). Minireview: Endocrine disruptors: Past lessons and future directions. *Molecular Endocrinology, 30*(8), 833–847.

92. Sies, H., Stahl, W., & Sevanian, A. (2005). Nutritional, dietary, and postprandial oxidative stress. *The Journal of Nutrition, 135*(5), 969–972.

93. Slizewska, G. (2020). What are lipids? Structure & function. *Expii*. Retrieved from: https://www.expii.com/t/what-are-lipids-structure-function-10071 (accessed on20 August 2024).

94. Soetan, K. O., Olaiya, C. O., & Oyewole, O. E. (2010). The importance of mineral elements for humans, domestic animals, and plants: A review. *African Journal of Food Science, 4*(5), 200–222.

95. Somero, G. N. (2012). The physiology of global change: Linking patterns to mechanisms. *Annual Review of Marine Science, 4*(1), 39–61.

96. Sordillo, L. M., & Aitken, S. L. (2009). Impact of oxidative stress on the health and immune function of dairy cattle. *Veterinary Immunology and Immunopathology, 128*(1–3), 104–109.

97. Speakman, J. R. (2005). Body size, energy metabolism, and lifespan. *Journal of Experimental Biology, 208*(9), 1717–1730.

98. Spears, J. W., & Weiss, W. P. (2008). Role of antioxidants and trace elements in health and immunity of transition dairy cows. *The Veterinary Journal, 176*(1), 70–76.

99. Squires, E. J. (2024). *Applied Animal Endocrinology*. CABI.

100. Studyadda. (2020). JEE Main, JEE Advanced, CBSE, NEET, IIT, free study packages, test papers, counselling, ask experts. *Studyadda*. Retrieved from: https://www.studyadda.com/notes/3rd-class/science/community-eating/types-of-animals-on-the-basis-of-food-habits/6294 (accessed on

20 August 2024).

101. Sulaiman, A. A., & Hussain, S. A. (2011). Protective effect of anthocyanins against chemicals-induced hemoglobin oxidation and erythrocytes membrane. *ResearchGate*. https://doi.org/10.13140/2.1.3217.8240.
102. Sundrum, A. (2015). Metabolic disorders in the transition period indicate that the dairy cows' ability to adapt is overstressed. *Animals, 5*(4), 978–1020.
103. Surai, N., Kochish, N., Fisinin, N., & Kidd, N. (2019). Antioxidant defence systems and oxidative stress in poultry biology: An update. *Antioxidants, 8*(7), 235. https://doi.org/10.3390/antiox8070235.
104. Tait-Burkard, C., Doeschl-Wilson, A., McGrew, M. J., Archibald, A. L., Sang, H. M., Houston, R. D., Whitelaw, C. B., & Watson, M. (2018). Livestock 2.0–Genome editing for fitter, healthier, and more productive farmed animals. *Genome Biology, 19*(1). https://doi.org/10.1186/s13059-018-1583-1.
105. Tu, Y., Chen, C., Pan, J., Xu, J., Zhou, Z., & Wang, C. Y. (2012). The ubiquitin proteasome pathway (UPP) in the regulation of cell cycle control and DNA damage repair and its implication in tumorigenesis. *Semantic Scholar*. https://www.semanticscholar.org/paper/The-Ubiquitin-Proteasome-Pathway-(UPP)-in-the-of-in-Tu-Chen/abdbb6b0df8c0bae5a42414aae0e23bc9ec79589 (accessed on 20 August 2024).
106. Usman, I. (2024). Animal biochemistry research topics. *Masstopics.com*. https://masstopics.com/research-paper/animal-biochemistry-research-topics/ (accessed on 20 August 2024).
107. Voet, D., & Voet, J. G. (2010). *Biochemistry*. John Wiley & Sons.
108. Wakil, S. J., & Abu-Elheiga, L. A. (2009). Fatty acid metabolism: Target for metabolic syndrome. *Journal of Lipid Research, 50*, S138–S143.
109. Wang, W. W., Wu, Z., Dai, Z., Yang, Y., Wang, J., & Wu, G. (2013). Glycine metabolism in animals and humans: Implications for nutrition and health.
110. Wang, Y., Gao, Y., Xia, C., Zhang, H., Qian, W., & Cao, Y. (2016). Pathway analysis of plasma different metabolites for dairy cow ketosis. *Italian Journal of Animal Science, 15*(3), 545–551. https://doi.org/10.1080/1828051x.2016.1180643.
111. Washington, I. M., & Van Hoosier, G. (2012). Clinical biochemistry and hematology. In *The Laboratory Rabbit, Guinea Pig, Hamster, and Other Rodents* (pp. 57–116). Elsevier.

112. Webster, K. (2003). Evolution of the coordinate regulation of glycolytic enzymes by hypoxia. *Journal of Experimental Biology, 206*(Pt 17), 2911–2922. https://doi.org/10.1242/jeb.00516.
113. Widmalm, G. (2013). A perspective on the primary and three-dimensional structures of carbohydrates. *Carbohydrate Research, 378*, 123–132.
114. Willmer, P., Stone, G., & Johnston, I. (2009). *Environmental Physiology of Animals*. John Wiley & Sons.
115. Worthington Biochemical. (2024). Chemical nature of enzymes. Retrieved from: https://www.worthington-biochem.com/tools-resources/intro-to-enzymes/chemical-nature-enzymes (accessed on 20 August 2024).
116. Wu, G., Bazer, F. W., Dai, Z., Li, D., Wang, J., & Wu, Z. (2014). Amino acid nutrition in animals: Protein synthesis and beyond. *Annual Review of Animal Biosciences, 2*(1), 387–417.
117. Yang, J., & Zhang, Y. (2015). I-TASSER server: New development for protein structure and function predictions. *Nucleic Acids Research, 43*(W1), W174–W181.
118. Zaitsev, S. Y., Bogolyubova, N. V., Zhang, X., & Brenig, B. (2020). Biochemical parameters, dynamic tensiometry, and circulating nucleic acids for cattle blood analysis: A review. *PeerJ, 8*, e8997.

INDEX

B

C

D

G

H

I

K

N

O

P

R

S

T

FSC
www.fsc.org
MIX
Papier | Fördert gute Waldnutzung
FSC® C083411

Kolibri 360 GmbH
Ferdinand-Jühlke-Straße 7
99095 Erfurt, Deutschland
produktsicherheit@kolibri360.de